Encyclopaedia of Animal Cell and Tissue Culture

Encyclopaedia of Animal Cell and Tissue Culture

Volume 3

Editor:
Dr. Anil M. Mane

AGROTECH PRESS
JAIPUR ▪ NEW DELHI

AGROTECH PRESS

Regd. Off. "Bony Residency", Gate No. 2
Opp. Tilak Public School, Vishweswariya Nagar,
Gopalpura Road, Jaipur - 302018, India
Ph: 0141 - 2761280, Telefax: 2761381
E-mail : agrotechpress@gmail.com

Delhi Office: 102, 1st Floor, Satyam House,
4327/3, Ansari Road, Darya Ganj,
New Delhi - 110002, India
Ph: 011-45652440
visit us: www.agrotechpress.com

Distributed By
KEYA INTERNATIONAL PRESS LLP
E-mail: keyainternationalpressllp@gmail.com

Published in India, 2025

ISBN : 9789384568276 (3 Vols. Set)

Typeset by Thomson Press

Printed and bound in India by Thomson Press Ltd.,
New Delhi

Contents

Preface

Animal biotechnology is the use of science and engineering to modify living organisms. The goal is to make products, to improve animals and to develop microorganisms for specific agricultural uses. Examples of animal biotechnology include creating transgenic animals (animals with one or more genes introduced by human intervention), using gene knock out technology to make animals with a specific inactivated gene and producing nearly identical animals by somatic cell nuclear transfer. The animal biotechnology in use today is built on a long history. Some of the first biotechnology in use includes traditional breeding techniques that date back to 5000 B.C.E. Such techniques include crossing diverse strains of animals to produce greater genetic variety. The offspring from these crosses then are bred selectively to produce the greatest number of desirable traits. For example, female horses have been bred with male donkeys to produce mules, and male horses have been bred with female donkeys to produce hinnies, for use as work animals, for the past 3,000 years. This method continues to be used today. The modern era of biotechnology began in 1953, when American biochemist James Watson and British biophysicist Francis Crick presented their double-helix model of DNA. That was followed by Swiss microbiologist Werner Arber's discovery in the 1960s of special enzymes, called restriction enzymes, in bacteria. These enzymes cut the DNA strands of any organism at precise points. In 1973, American geneticist Stanley Cohen and American biochemist Herbert Boyer removed a specific gene from one bacterium and inserted it into another using restriction enzymes. That event marked the beginning of recombinant DNA technology, or genetic engineering. In 1977, genes from other organisms were transferred to bacteria, an achievement that led eventually to the first transfer of a human gene. Animal biotechnology in use today is based on the science of genetic engineering. Under the umbrella of genetic engineering exist other technologies, such as transgenics and cloning, that also are used in animal biotechnology. Transgenics is the transferal of a specific gene from one organism to another. Gene splicing is used to introduce one or more genes of an organism into a second organism.

A transgenic animal is created once the second organism incorporates the new DNA into its own genetic material. In gene splicing, DNA cannot be transferred directly from its original organism, the donor, to the recipient organism, or the host. Instead, the donor DNA must be cut and pasted, or recombined, into a compatible fragment of DNA from a vector — an organism that can carry the donor DNA into the host. The host organism often is a rapidly multiplying microorganism such as a harmless bacterium, which serves as a factory where the recombined DNA can be duplicated in large quantities. The subsequently produced protein then can be removed from the host and used as a genetically engineered product in humans, other animals, plants, bacteria or viruses. The donor DNA can be introduced directly into an organism by techniques such as injection through the cell walls of plants or into the fertilized egg of an animal. This transferring of genes alters the characteristics of the organism by changing its protein makeup. Proteins, including enzymes and hormones, perform many vital functions in organisms. Individual genes direct an animal's characteristics through the production of proteins. Scientists use reproductive cloning techniques to produce multiple copies of mammals that are nearly identical copies of other animals, including transgenic animals, genetically superior animals and animals that produce high quantities of milk or have some other desirable trait. To date, cattle, sheep, pigs, goats, horses, mules, cats, rats and mice have been cloned, beginning with the first cloned animal, a sheep named Dolly, in 1996.

Animal biotechnology has many potential uses. Since the early 1980s, transgenic animals have been created with increased growth rates, enhanced lean muscle mass, enhanced resistance to disease or improved use of dietary phosphorous to lessen the environmental impacts of animal manure. Transgenic poultry, swine, goats and cattle that generate large quantities of human proteins in eggs, milk, blood or urine also have been produced, with the goal of using these products as human pharmaceuticals. Human pharmaceutical proteins include enzymes, clotting factors, albumin and antibodies. The major factor limiting the widespread use of transgenic animals in agricultural production systems is their relatively inefficient production rate.

This book extensively deals with the basics and recent developments in Animal Biotechnology in a simplified manner. The detailed and systematic discussion about each topic with illustrations and examples would help the readers to understand the different aspects of Animal Biotechnology. The book also covers the Intellectual Property Rights, and biosafety guidelines to be adopted in the laboratory are also included. This meticulously written book would be of immense help to students, researchers and teachers in the field of biotechnology.

—Editor

1

Future Growth and Directions

The proliferation of bioinformatics tools for genetic analysis aids researchers in identifying and categorizing genes and gene sets of interest in their work; however, the large variety of tools that perform substantially similar aggregative and analytical functions can also confuse and frustrate new users.

The decentralization encouraged by aggregative tools allows individual research groups to maintain specialized servers dedicated to specific types of data analysis in the expectation that their output will be collected into a larger report on a gene or protein of interest to other researchers.

Data produced by microarray experiments, two-hybrid screening, and other high-throughput biological experiments is voluminous and difficult to analyse by hand; the efforts of structural genomics collaborations that are aimed at quickly solving large numbers of highly varied protein structures also increase the need for integration between sequence and structure databases and portals. This impetus towar developing more comprehensive and more user-friendly methods of sequence profiling makes this an active area of research among current genomics researchers.

Genetically Modified Organism

A genetically modified organism (GMO) or genetically engineered organism (GEO) is an organism whose genetic material has been altered using genetic engineering techniques. These techniques, generally known as recombinant DNA technology, use DNA molecules from different sources, which are combined into one molecule to create a new set of genes. This DNA is then transferred into an organism, giving it modified or novel genes. Transgenic organisms, a subset of GMOs, are organisms which have inserted DNA that originated in a different species.

Production

Genetic modification involves the insertion or deletion of genes. When genes are inserted, they usually come from a different species, which is a form of horizontal gene transfer.

Figure: *GloFish, the first genetically modified animal to be sold as a pet*

In nature this can occur when exogenous DNA penetrates the cell membrane for any reason.

To do this artificially may require attaching the genes to a virus or just physically inserting the extra DNA into the nucleus of the intended host with a very small syringe, or with very small particles fired from a gene gun. However, other methods exploit natural forms of gene transfer, such as the ability of Agrobacterium to transfer genetic material to plants, or the ability of lentiviruses to transfer genes to animal cells.

History

The general principle of producing a GMO is to add new genetic material into an organism's genome. This is called genetic engineering and was made possible through the discovery of DNA and the creation of the first recombinant bacteria in 1973; an existing bacterium E. coli expressing an exogenic Salmonella gene. This led to concerns in the scientific community about potential risks from genetic engineering, which were first discussed in depth at the Asilomar Conference in 1975. One of the main recommendations from this meeting was that government oversight of recombinant DNA research should be established until the technology was deemed safe. Herbert Boyer then founded the first company to use recombinant DNA technology, Genentech, and in 1978 the company announced creation of an E. coli strain producing the human protein insulin.

In 1986, field tests of bacteria genetically engineered to protect plants from frost damage (ice-minus bacteria) at a small biotechnology company called Advanced Genetic Sciences of Oakland, California, were repeatedly delayed by opponents of biotechnology. In the same year, a proposed field test of a microbe genetically engineered for a pest resistance protein by Monsanto Company was dropped.

In the late 1980s and early 1990s, guidance on assessing the safety of genetically engineered plants and food emerged from organizations including the FAO and WHO.

Small scale experimental plantings of genetically modified (GM) plants began in Canada and the U.S. in the late 1980s. The first approvals for large scale, commercial cultivation came in the mid 1990s. Since that time, adoption of GM plants by farmers has increased annually.

Uses

GMOs are used in biological and medical research, production of pharmaceutical drugs, experimental medicine (e.g. gene therapy), and agriculture (e.g. golden rice). The term "genetically modified organism" does not always imply, but can include, targeted insertions of genes from one species into another. For example, a gene from a jellyfish, encoding a fluorescent protein called GFP, can be physically linked and thus co-expressed with mammalian genes to identify the location of the protein encoded by the GFP-tagged gene in the mammalian cell. Such methods are useful tools for biologists in many areas of research, including those who study the mechanisms of human and other diseases or fundamental biological processes in eukaryotic or prokaryotic cells.

To date the most controversial but also the most widely adopted application of GMO technology is patent-protected food crops which are resistant to commercial herbicides or are able to produce pesticidal proteins from within the plant, or stacked trait seeds, which do both. The largest share of the GMO crops planted globally are owned by the US firm Monsanto. In 2007, Monsanto's trait technologies were planted on 246 million acres (1,000,000 km) throughout the world, a growth of 13 percent from 2006. However, patents on the first Monsanto products to enter the marketplace will begin to expire in 2014, democratizing Monsanto products. In addition, a 2007 report from the European Joint Research Commission predicts that by 2015, more than 40 per cent of new GM plants entering the global marketplace will have been developed in

Asia. In the corn market, Monsanto's triple-stack corn—which combines Roundup Ready 2 weed control technology with YieldGard Corn Borer and YieldGard Rootworm insect control—is the market leader in the United States. U.S. corn farmers planted more than 32 million acres (130,000 km) of triple-stack corn in 2008, and it is estimated the product could be planted on 56 million acres (230,000 km) in 2014–2015. In the cotton market, Bollgard II with Roundup Ready Flex was planted on approximately 5 million acres (20,000 km) of U.S. cotton in 2008.

According to the International Service for the Acquisition of Agri-Biotech Applications (ISAAA), in 2010 approximately 15 million farmers grew biotech crops in 29 countries. Over 90% of the farmers were resource-poor in developing countries. 6.5 million farmers in China and 6.3 million small farmers in India grew biotech crops (mostly Bacillus thuringiensis cotton).

The Philippines, South Africa (biotech cotton, maize and soybeans often grown by subsistence women farmers) and another twelve developing countries also grew biotech crops in 2009. 10 million more small and resource-poor farmers may have been secondary beneficiaries of Bt cotton in China.

The global commercial value of biotech crops grown in 2008 was estimated to be US$130 billion.

In the United States, the United States Department of Agriculture (USDA) reports on the total area of GMO varieties planted. According to National Agricultural Statistics Service, the states published in these tables represent 81–86 percent of all corn planted area, 88–90 percent of all soybean planted area, and 81–93 percent of all upland cotton planted area (depending on the year).

USDA does not collect data for global area. Estimates are produced by the International Service for the Acquisition of Agri-biotech Applications (ISAAA) and can be found in the report, "Global Status of Commercialized Transgenic Crops: 2007".

Transgenic animals are also becoming useful commercially. On February 6, 2009 the U.S. Food and Drug Administration approved the first human biological drug produced from such an animal, a goat. The drug, ATryn, is an anticoagulant which reduces the probability of blood clots during surgery or childbirth. It is extracted from the goat's milk.

Detection

Testing on GMOs in food and feed is routinely done by molecular techniques like DNA microarrays or qPCR. The test can be based on screening elements (like p35S, tNos, pat, or bar) or event-specific markers for the official GMOs (like Mon810, Bt11, or GT73). The array-based method combines multiplex

PCR and array technology to screen samples for different potential GMOs, combining different approaches (screening elements, plant-specific markers, and event-specific markers).

The qPCR is used to detect specific GMO events by usage of specific primers for screening elements or event-specific markers.

To avoid any kind of false positive or false negative testing outcome, comprehensive controls for every step of the process is mandatory. A CaMV check is important to avoid false positive outcomes based on virus contamination of the sample.

Transgenic Microbes

Bacteria were the first organisms to be modified in the laboratory, due to their simple genetics. These organisms are now used for several purposes, and are particularly important in producing large amounts of pure human proteins for use in medicine.

Genetically modified bacteria are used to produce the protein insulin to treat diabetes. Similar bacteria have been used to produce clotting factors to treat haemophilia, and human growth hormone to treat various forms of dwarfism.

Transgenic Animals

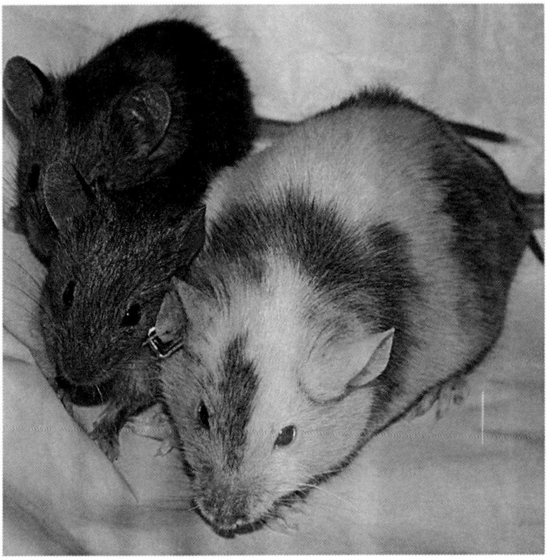

Figure: *Some chimeras, like the blotched mouse shown, are created through genetic modification techniques like gene targeting.*

Transgenic animals are used as experimental models to perform phenotypic and for testing in biomedical research.

Genetically Modified (Genetically Engineered) animals are becoming more vital to the discovery and development of cures and treatments for many serious diseases. By altering the DNA or transferring DNA to an animal, we can develop certain proteins that may be used in medical treatment. Stable expressions of human proteins have been developed in many animals, including sheep, pigs, and rats.

Some examples are: Human-alpha-1-antitrypsin, which has been developed in sheep and is used in treating humans with this deficency and transgenic pigs with human-histo-compatibility have been studied in the hopes that the organs will be suitable for transplant with less chances of rejection.

Transgenic livestock have been used as bioreactors since the 1990s. Many medicines, including insulin and many immunizations are developed in transgenic animals. In March 2011, the bioactive recombinant Human Lysozyme was expressed in the milk of cloned transgenic cattle. This field is growing rapidly and new pharming uses are being discovered and developed. The extent that trangenic animals will be useful in the medical field as well as other fields is very promising based on results thus far.

Fruit Flies

In biological research, transgenic fruit flies (Drosophila melanogaster) are model organisms used to study the effects of genetic changes on development. Fruit flies are often preferred over other animals due to their short life cycle, low maintenance requirements, and relatively simple genome compared to many vertebrates.

Mosquitoes

In 2010, scientists created "malaria-resistant mosquitoes" in the laboratory. The World Health Organisation estimated that Malaria killed almost one million people in 2008. Genetically modified male mosquitoes containing a lethal gene have been developed in order to combat the spread of Dengue fever.

Aedes aegypti mosquitoes, the single most important carrier of dengue fever, were reduced by 80% in a 2010 trial of these GM mosquitoes in the Cayman Islands. Between 50 - 100 million people are affected by Dengue fever every year and 40,000 people die from it.

Bollworms

A strain of Pectinophora gossypiella (Pink bollworm) has been developed that contains a fluorescent marker in their DNA. This allows researchers to monitor bollworms that have been sterilized by radiation and released in order to reduce bollworm infestation.

Mammals

Genetically modified mammals are an important category of genetically modified organisms. Transgenic mice are often used to study cellular and tissue-specific responses to disease.

In 1999, scientists at the University of Guelph in Ontario, Canada created the genetically engineered Enviropig. The Enviropig excretes from 30 to 70.7% less phosphorus in manure depending upon the age and diet. In February 2010, Environment Canada determined that Enviropigs are in compliance with the Canadian Environmental Protection Act and can be produced outside of the research context in controlled facilities where they are segregated from other animals.

In 2009, scientists in Japan announced that they had successfully transferred a gene into a primate species (marmosets) and produced a stable line of breeding transgenic primates for the first time. Their first research target for these marmosets was Parkinson's disease, but they were also considering Amyotrophic lateral sclerosis and Huntington's disease.

In 2011, scientists in China released news that they have introduced human genes into 300 dairy cows to produce milk with the same properties as human breast milk. Aside from milk production, the researchers claim these transgenic cows to be identical to regular cows.

Cnidarians

Cnidarians such as Hydra and the sea anemone Nematostella vectensis have become attractive model organisms to study the evolution of immunity and certain developmental processes. An important technical breakthrough was the development of procedures for generation of stably transgenic hydras and sea anemones by embryo microinjection.

Fish

Genetically modified fish have promoters driving an over-production of "all fish" growth hormone. This resulted in dramatic growth enhancement in several species, including salmonids, carps and tilapias.

Gene Therapy

Gene therapy, uses genetically modified viruses to deliver genes that can cure disease into human cells. Although gene therapy is still relatively new, it has had some successes. It has been used to treat genetic disorders such as severe combined immunodeficiency, and treatments are being developed for a range of other currently incurable diseases, such as cystic fibrosis, sickle cell anemia, Parkinson's disease and muscular dystrophy. Current gene therapy technology only targets the non-reproductive cells meaning that any changes introduced by the treatment can not be transmitted to the next generation. Gene therapy targeting the reproductive cells—so-called "Germ line Gene Therapy"—is very controversial and is unlikely to be developed in the near future.

Transgenic Plants

Transgenic plants have been engineered to possess several desirable traits, such as resistance to pests, herbicides, or harsh environmental conditions, improved product shelf life, and increased nutritional value. Since the first commermmodified to be tolerant to the herbicides glufosinate and glyphosate, to be resistant to virus damage as in Ringspot virus-resistant GM papaya, grown in Hawaii, and to produce the Bt toxin, an insecticide that is non-toxic to mammals. Most GM crops grown today have been modified with "input traits", which provide benefits mainly to farmers. The GM oilseed crops on the market today offer improved oil profiles for processing or healthier edible oils.

The GM crops in development offer a wider array of environmental and consumer benefits such as nutritional enhancement, drought and stress tolerance. GM plants are being developed by both private companies and public research institutions such as CIMMYT, the International Maize and Wheat Improvement Centre. Other examples include a genetically modified sweet potato, enhanced with protein and other nutrients, while golden rice, developed by the International Rice Research Institute (IRRI), has been discussed as a possible cure for Vitamin A deficiency.

Scientists at the University of York developed a weed (Arabidopsis thaliana) which contained genes from bacteria that can clean up TNT and RDX explosive contaminants from the soil and it was hoped that this weed would eliminate this pollution. 16 million hectares in the USA (1.5% of the total surface) are estimated to be contaminated with TNT and RDX. However the weed Arabidopsis thaliana was not tough enough to withstand the environment on military test grounds and research is continuing with the University of Washington to develop a tougher native grass.

The coexistence of GM plants with conventional and organic crops has raised significant concern in many European countries. Due to relatively high demand from European consumers for the freedom of choice between GM and non-GM foods, EU regulations require measures to avoid mixing of foods and feed produced from GM crops and conventional or organic crops. European research programs such as Co-Extra, Transcontainer, and SIGMEA are investigating appropriate tools and rules. At the field level, biological containment methods include isolation distance and pollen barriers. Such measures are generally not used in North America because they are very costly and there are no safety-related reasons to employ them.

Cisgenic Plants

Cisgenesis, sometimes also called Intragenesis, is a product designation for a category of genetically engineered plants. A variety of classification schemes have been proposed, that order genetically modified organisms based on the nature of introduced genotypical changes rather than the process of genetic engineering.

While some genetically modified plants are developed by the introduction of a gene originating from distant, sexually incompatible species into the host genome, cisgenic plants contain genes which have been isolated either directly from the host species or from sexually compatible species. The new genes are introduced using recombinant DNA methods and gene transfer. Some scientists hope that the approval process of cisgenic plants might be simpler than that of proper transgenics, but it remains to be seen.

Regulation

The USA is the largest commercial grower of genetically modified crops in the world. For a genetically modified organism to be approved for release it is assessed by the Animal and Plant Health Inspection Service (APHIS) agency within the United States Department of Agriculture (USDA), the Food and Drug Administration (FDA) and the Environmental Protection Agency (EPA).

The USDA evaluated the plants potential to become weeds, the FDA reviewed plants that could enter or alter the food supply and the EPA regulated the genetically modified plants with pesticide properties. Most developed genetically modified plants are reviewed by at least two of the agencies, with many subject to all three. Final approval can still be denied by individual counties within each state. In 2004, Mendocino County, California became the first and only county to impose a ban on the "Propagation, Cultivation, Raising, and Growing of Genetically Modified Organisms", the measure passing with a 57% majority.

The European Union (EU) has possibly the most stringent GMO regulations in the world. All GMOs, along with irradiated food, are considered "new food" and subject to extensive, case-by-case, science based food evaluation by the European Food Safety Authority (EFSA). The EFSA reports to the European Commission who then draft a proposal which if accepted will be adopted by the EC or passed on to the Council of Agricultural Ministers. There is also a safeguard clause that Member States can invoke to restrict or prohibit the use and/or sale of a GMO within their territory if they have a justifiable reasons to consider that the approved GMO constitutes a risk to human health or the environment. In February 2008 the French government used the safeguard clause to ban the cultivation of MON810 after Senator Jean-François Le Grand, chairman of a committee set up to evaluate biotechnology, said there were "serious doubts" about the safety of the product. In April 2009 German Federal Minister Ilse Aigner announced an immediate halt to cultivation and marketing of MON810 maize under the safeguard clause.

Currently (2010) the only GMO food crop with approval for cultivation in Europe is the GM maize MON810, which gained approval in 1998. On 2 March 2010 a second GMO, a potato called Amflora, was approved for cultivation for industrial applications in the EU by the European Commission and was grown in Germany, Sweden and the Czech Republic that year. Co-existence of GM and non-GM crops is regulated by the use of buffer zones and isolation distances between the GM and non-GM crops. The regulations concerning the import and sale of GMOs for human and animal consumption grown outside the EU involve providing freedom of choice to the farmers and consumers. Twice GMOs unapproved by the EC have arrived in the EU and been forced to return to their port of origin. The first was in 2006 when a shipment of rice from America containing an experimental GMO variety (LLRice601) not meant for commercialisation arrived at Rotterdam. The second in 2009 when trace amounts of a GMZO maize approved in the US were found in a "non-GM" soy flour cargo.

Genetic engineering in Australia is overseen by the Office of the Gene Technology Regulator (OGTR), a Commonwealth Government Authority within the Department of Health and Ageing. The OGTR reports directly to Parliament through a Ministerial Council on Gene Technology and has legislative powers. The OGTR decides on license applications for the release of all genetically modified organisms and Food Standards Australia New Zealand regulates any GM food. Individual state governments are able to assess the impact of release on markets and trade and apply further legislation to control approved genetically modified productsma for commercial cultivation in 2008. GM canola is grown in Western Australia, while South Australia and Tasmania have extended their moratoriums on all genetically modified crops.

Health Canada, under the Food and Drugs Act, and the Canadian Food Inspection Agency are responsible for evaluating the safety and nutritional value of genetically modified foods. The committee that reviewed the regulations in 2003 was accused by environmental and citizen groups of not representing the full spectrum of public interests and for being too closely aligned to industry groups. In central and South America Mexico, Honduras, Costa Rico, Colombia, Bolivia, Paraguay, Chile, Argentina, Uruguay and Brazil all grow GM crops. In Argentina the National Agricultural Biotechnology Advisory Committee (environmental impact), the National Service of Health and Agrifood Quality (food safety) and the National Agribusiness Direction (effect on trade) assess GM products for release, with the final decision made by the Secretariat of Agriculture, Livestock, Fishery and Food. In Brazil the National Biosafety Technical Commission is responsible for assessing environmental and food safety and prepares guidelines for transport, importation and field experiments involving GM products.

The Council of Ministers evaluates the commercial and economical issues with release. Mexico's senate passed a law allowing planting and selling of genetically modified cotton and soybean in Mexicoin 2005 and in 2009 the government enacted statutory provisions for the regulation of genetically modified maize Mexico is the centre of diversity for maize and concerns have been raised about the impact genetically modified maize could have on local strains.

GM crops in China go through three phases of field trials (pilot field testing, environmental release testing, and preproduction testing) before they are submitted to the Office of Agricultural Genetic Engineering Biosafety Administration (OAGEBA) for assessment. Producers must apply to OAGEBA at each stage of the field tests. The Chinese Ministry of Science and Technology developed the first biosafety regulations for GM products in 1993 and they were updated in 2001. Most of the National Biosafety Committee are involved in biotechnology leading to criticisms that they do not represent a wide enough range of public concerns. India regulators cleared the Bt brinjal, a genetically modified eggplant, for commercialization in October 2009. Following opposition from some scientists, farmers and environmental groups a moratorium was imposed on its release in February 2010. The only other Asian country to currently grow GM crops is the Phillipenes.

In 2010, the Common Market for Eastern and Southern Africa (COMESA) proposed a policy where new GM crops would be scientifically assessed by COMESA. If it was demeed safe for the environmental and human health permission would be granted for the crop to be grown in all 19 member countries, although the final decision would be left to each individual country. In 2010 South Africa was the major grower of genetically modified crops in Africa, with smaller areas planted in Burkina Faso and Egypt. Burkina Faso has established a National Biosafety Agency that regulates GM products with advice from various governmental and non-governmental advisory committees.

Kenya passed laws in 2011 which allowed the production and importation of GM crops. The Zambian government rejected a consignment of GMO maize supplied by donors during a famine in 2002 on the basis of the Cartagena Protocol.

One of the key issues concerning regulators is whether GM products should be labeled. Labeling can be mandatory up to a threshold GM content level (which varies between countries) or voluntary. A study investigating voluntary labeling in South Africa found that 31% of products labeled as GMO-free had a GM content above 1.0%. In Canada and the USA labeling of GM food is voluntary, while in Europe it all food (including processed food) or feed which contains greater than 0.9% of approved GMOs must be labelled.

Controversy

The examples and perspective in this article may not represent a worldwide view of the subject. Please improve this article and discuss the issue on the talk page.

Biological Process

The use of genetically modified organisms has sparked significant controversy in many areas. Some groups or individuals see the generation and use of GMO as intolerable meddling with biological states or processes that have naturally evolved over long periods of time, while others are concerned about the limitations of modern science to fully comprehend all of the potential negative ramifications of genetic manipulation. Other people see this as a continuation in the role humanity has occupied for thousands of years, modifying the genetics of crops by selecting specimen of crops with the most desirable characteristics as parent for the next generation of crops.

Foodchain

The safety of GMOs in the foodchain has been questioned by some environmental groups, with concerns such as the possibilities that GMOs could introduce new allergens into foods, or contribute to the spread of antibiotic resistance. According to a study published in 1999, there was no current evidence to suggest that the processes used to genetically modify food were inherently harmful. However, a number of more recent studies have raised concern, and environmental groups still discourage consumption in many countries, claiming that GM foods are unnatural and therefore unsafe. Such concerns have led to the adoption of laws and regulations that require safety testing of any new organism produced for human consumption.

GMOs' proponents note that because of the safety testing requirements imposed on GM foods, the risk of introducing a plant variety with a new allergen or toxin using genetic modification is much smaller than using traditional breeding processes. Transgenesis has less impact on the expression of genomes or on protein and metabolite levels than conventional breeding or plant (non-directed) mutagenesis. An example of an allergenic plant created using traditional breeding is the kiwi. One article calculated that the marketing of GM salmon could reduce the cost of salmon by half, thus increasing salmon consumption and preventing 1,400 deaths from heart attack a year in the United States.

Trade in Europe and Africa

In response to negative public opinion, Monsanto announced its decision to remove their seed cereal business from Europe, and environmentalists crashed a World Trade Organization conference in Cancun that promoted GM foods and was sponsored by Committee for a Constructive Tomorrow (CFACT). Some African nations have refused emergency food aid from developed countries, fearing that the food is unsafe. During a conference in the Ethiopian capital of Addis Ababa, Kingsley Amoako, Executive Secretary of the United Nations Economic Commission for Africa (UNECA), encouraged African nations to accept genetically modified food and expressed dissatisfaction in the public's negative opinion of biotechnology.

Agricultural Surpluses

Patrick Mulvany, Chairman of the UK Food Group, accused some governments, especially the Bush administration, of using GM food aid as a way to dispose of unwanted agricultural surpluses. The UN blamed food companies and accused them of violating human rights, calling on governments to regulate these profit-driven firms. It is widely believed that the acceptance of biotechnology and genetically modified foods will also benefit rich research companies and could possibly benefit them more than consumers in underdeveloped nations.

Labeling

While some groups advocate the complete prohibition of GMOs, others call for mandatory labeling of genetically modified food or other products. Other controversies include the definition of patent and property pertaining to products of genetic engineering. According to the documentary Food, Inc. efforts to introduce labeling of GMOs has repeatedly met resistance from lobbyists and politicians affiliated with companies like Monsanto.

Testing

Bruce Stutz's article, "Wanted: GM Seeds for Study," highlights a story of two dozen scientists who spoke out against the research restrictions put forth by companies producing genetically modified (GM) seeds such as DuPont, Monsanto, and Syngenta. In February 2009, after scientists warned the U.S. Environmental Protection Agency (EPA) "that industry influence had made independent analyses of transgenic crops impossible," the American Seed Trade Association (ASTA) agreed that they "would allow researchers greater freedom to study the effects of GM food crops." This agreement left many scientists optimistic about the future, but there is little optimism as to whether this agreement has the ability to "alter what has been a research environment rife with obstruction and suspicion."

Impoverished Nations

Some groups believe that impoverished nations will not reap the benefits of biotechnology because they do not have easy access to these developments, cannot afford modern agricultural equipment, and certain aspects of the system revolving around intellectual property rights are unfair to "undeveloped countries". For example, The CGIAR (Consultative Group of International Agricultural Research) is an aid and research organization that has been working to achieve sustainable food security and decrease poverty in undeveloped countries since its formation in 1971. In an evaluation of CGIAR, the World Bank praised its efforts but suggested a shift to genetics research and productivity enhancement. This plan has several obstacles such as patents, commercial licenses, and the difficulty that third world countries have in accessing the international collection of genetic resources and other intellectual property rights that would educate them about modern technology. The International Treaty on Plant Genetic Resources for Food and Agriculture has attempted to remedy this problem, but results have been inconsistent. As a result, "orphan crops", such as teff, millets, cowpeas, and indigenous plants, are important in the countries where they are grown, but receive little investment.

Private Investments

The development and implementation of policies is designed to encourage private investments in research and marketing biotechnology that will meet the needs of poverty-stricken nations, increased research on other problems faced by poor nations, and joint efforts by the public and private sectors to ensure the efficient use of technology developed by industrialized nations have been suggested. In addition, industrialized nations have not tested GM technology on tropical plants, focusing on those that grow in temperate climates, even though undeveloped nations and the people that need the extra food live primarily in tropical climates. Some European scientists are concerned that political factors and ideology prevent unbiased assessment of GM technology in some EU countries, with a negative effect on the whole community.

Transgenic Organisms

Another important controversy is the possibility of unforeseen local and global effects as a result of transgenic organisms proliferating. The basic ethical issues involved in genetic research are discussed in the article on genetic engineering. Some critics have raised the concern that conventionally-bred crop plants can be cross-pollinated (bred) from the pollen of modified plants. Pollen can be dispersed over large areas by wind, animals, and insects. In 2007, the U.S. Department of Agriculture fined Scotts Miracle-Gro $500,000 when modified genetic material from creeping bentgrass, a new golf-course grass Scotts had been testing, was found within close relatives of the same genus (Agrostis) as well as in native grasses up to 21 km (13 mi) away from the test sites, released when freshly cut grass was blown by the wind.

GM proponents point out that outcrossing, as this process is known, is not new. The same thing happens with any new open-pollinated crop variety—newly introduced traits can potentially cross out into neighboring crop plants of the same species and, in some cases, to closely related wild relatives. Defenders of GM technology point out that each GM crop is assessed on a case-by-case basis to determine if there is any risk associated with the outcrossing of the GM trait into wild plant populations.

The fact that a GM plant may outcross with a related wild relative is not, in itself, a risk unless such an occurrence has negative consequences. If, for example, an herbicide resistance trait was to cross into a wild relative of a crop plant it can be predicted that this would not have any consequences except in areas where herbicides are sprayed, such as a farm. In such a setting the farmer can manage this risk by rotating herbicides.

The European Union funds research programes such as Co-Extra, that investigate options and technologies on the coexistence of GM and conventional farming. This also includes research on biological containment strategies and other measures to prevent outcrossing and enable the implementation of coexistence.

If patented genes are outcrossed, even accidentally, to other commercial fields and a person deliberately selects the outcrossed plants for subsequent planting then the patent holder has the right to control the use of those crops. This was supported in Canadian law in the case of Monsanto Canada Inc. v. Schmeiser.

"Terminator" and "Traitor"

An often cited controversy is a "Technology Protection" technology dubbed 'Terminator'. This uncommercialized technology would allow the production of first generation crops that would not generate seeds in the second generation because the plants yield sterile seeds. The patent for this so-called "terminator" gene technology is owned by Delta and Pine Land Company and the United States Department of Agriculture. Delta and Pine Land was bought by Monsanto Company in August 2006. Similarly, the hypothetical trait-specific Genetic Use Restriction Technology, also known as 'Traitor' or 'T-GURT', requires application of a chemical to genetically modified crops to reactivate engineered traits. This technology is intended both to limit the spread of genetically engineered plants, and to require farmers to pay yearly to reactivate the genetically engineered traits of their crops. Genetic Use Restriction Technology is under development by companies including Monsanto and AstraZeneca.

In addition to the commercial protection of proprietary technology in self-pollinating crops such as soybean (a generally contentious issue), another purpose of the terminator gene is to prevent the escape of genetically modified traits from cross-pollinating crops into wild-type species by sterilizing any resultant hybrids. Some environmentalist groups, while considering outcrossing of GM plants dangerous, felt the technology would prevent re-use of seed by farmers growing such terminator varieties in the developing world and was ostensibly a means to exercise patent claims. However other environmental groups welcomed the terminator gene as a means of preventing GM crops from mixing with natural crops.

Hybrid seeds were commonly used in the developed countries long before the introduction of GM crops. Some hybrid crop seeds cannot be saved, so purchasing new seed every year is already a standard agricultural practice for a majority of farms.

There are technologies evolving which contain the transgene by biological means and still can provide fertile seeds using fertility restorer functions. Such methods are being developed by several EU research programes, among them Transcontainer and Co-Extra.

BioSteel

BioSteel is a trademark name for a high-strength based fibre material made of the recombinant spider silk-like protein extracted from the milk of transgenic goats, made by Nexia Biotechnologies.

The company has created lines of goats that produce recombinant versions of either the MaSpI or MaSpII dragline silk proteins in their milk. When the female goats lactate, the milk, containing the recombinant silk, is harvested and subjected to chromatographic techniques to purify the recombinant silk proteins.

The purified silk proteins are then dried, dissolved using solvents (DOPE formation) and transformed into microfibers using wet-spinning fibre production methodologies. The spun fibres so far have tenacities in the range of 2 - 3 grams/denier and elongation range of 25-45%. The "Biosteel biopolymer" has been transformed into nanofibres and nano-meshes using the electrospinning technique.

Biosteel and other biopolymers are being researched to provide lightweight, strong, and versatile materials for a variety of medical and industrial applications. Nexia Biotechnologies plans to use the spider silk from the milk of transgenic goats for bulletproof vests and anti-ballistic missile systems.

Unfortunately, no one has been able to produce it in commercial quantities. Nexia is the only company which has successfully produced fibres from recombinant spider silk and is currently in the process of developing commercial quantities of BioSteel using its transgenic goat technology.

Chimera (genetics)

A chimera or chimaera is a single organism (usually an animal) that is composed of two or more different populations of genetically distinct cells that originated from different zygotes involved in sexual reproduction. If the different cells have emerged from the same zygote, the organism is called a mosaic. Chimeras are formed from four parent cells (two fertilized eggs or early embryos fused together). Each population of cells keeps its own character and the resulting organism is a mixture of tissues. Chimeras are typically seen in animals; there are some reports on human chimerism. Plant chimeras are discussed separately.

This condition is either inherited, or it is acquired through the infusion of allogeneic hematopoietic cells during transplantation or transfusion. In nonidentical twins, chimerism occurs by means of blood-vessel anastomoses. The likelihood of offspring being a chimera is increased if it is created via in vitro fertilization. Chimeras can often breed, but the fertility and type of offspring depends on which cell line gave rise to the ovaries or testes; varying degrees of intersexuality may result if one set of cells is genetically female and another genetically male.

Tetragametic Chimerism

Tetragametic chimerism is a form of congenital chimerism. This condition occurs through the fertilization of two separate ova by two sperm, followed by the fusion of the two at the blastocyst or zygote stages. This results in the development of an organism with intermingled cell lines. Put another way, the chimera is formed from the merging of two nonidentical twins (although a similar merging presumably occurs with identical twins, but as their DNA is almost identical, the presence would not be immediately detectable in a very early (zygote or blastocyst) phase. As such, they can be male, female, or hermaphroditic.

As the organism develops, it can come to possess organs that have different sets of chromosomes. For example, the chimera may have a liver composed of cells with one set of chromosomes and have a kidney composed of cells with a second set of chromosomes. This has occurred in humans, and at one time was thought to be extremely rare, though more recent evidence suggests that it is not as rare as previously believed.

This is particularly true for the marmoset. Recent research shows most marmosets are chimerae, sharing DNA with their fraternal twins. Most chimerae will go through life without realizing they are chimeras. The difference in phenotypes may be subtle (e.g., having a hitchhiker's thumb and a straight thumb, eyes of slightly different colors, differential hair growth on opposite sides of the body, etc.) or completely undetectable. Chimera may also show, under a certain spectrum of UV light, distinctive marks on the back resembling that of arrow points pointing downwards from the shoulders down to the lower back; this is one expression of pigment unevenness called Blaschko's lines. Affected persons may be identified by the finding of two populations of red cells or, if the zygotes are of opposite sex, ambiguous genitalia and hermaphroditism alone or in combination; such persons sometimes also have patchy skin, hair, or eye pigmentation (heterochromia). If the blastocysts are of opposite sex, genitals of both sexes may be formed, either ovary and testis, or combined ovotestes, in one rare form of intersexuality, a condition previously known as true hermaphroditism.

Note that the frequency of this condition does not indicate the true prevalence of chimerism. Most chimerae composed of both male and female cells probably do not have an intersex condition, as might be expected if the two cell populations were evenly blended throughout the body. Often, most or all of the cells of a single cell type will be composed of a single cell line, i.e. The blood may be composed prominently of one cell line, and the internal organs of the other cell line. Genitalia produce the hormones responsible for other sex characteristics. If the sex organs are homogeneous, the individual will not be expected to exhibit any intersex traits.

Natural chimeras are almost never detected unless they exhibit abnormalities such as male/female or hermaphrodite characteristics or uneven skin pigmentation. The most noticeable are some male tortoiseshell cats or animals with ambiguous sex organs.

The existence of chimerism is problematic for DNA testing, a fact with implications for family and criminal law. The Lydia Fairchild case, for example, was brought to court after DNA testing apparently showed that her children could not be hers. Fraud charges were filed against her and her custody of her children was challenged. The charge against her was dismissed when it became clear that Lydia was a chimera, with the matching DNA being found in her cervical tissue. Another case was that of Karen Keegan, who was also in danger of losing her children, after a DNA test for a kidney transplant seemed to show she wasn't the mother of her children. The tetragametic state has important implications for organ or stem-cell transplantation. Chimeras typically have immunologic tolerance to both cell lines.

Microchimerism

Microchimerism is the presence of a small number of cells that are genetically distinct from those of the host individual. Apparently, this phenomenon is related to certain types of autoimmune diseases, however, the mechanisms responsible for this relationship are unclear.

"Parasitic" Chimerism in Anglerfish

Chimerism occurs naturally in adult Ceratioid anglerfish and is in fact a natural and essential part of their life cycle. Once a male is born, it begins its search for a female. Using strong olfactory glands, the male searches until it locates a female anglerfish. The male, less than an inch in length, bites into her skin and releases an enzyme that digests the skin of his mouth and her body, fusing the pair down to the blood-vessel level. While this attachment has become necessary for the male's survival, it will eventually consume him, as both anglerfish fuse into a single hermaphroditic individual. Sometimes in this odd ritual, more than one male will attach to a single female as a 'parasite'. They will both be consumed into the body of the larger female angler. Once fused to a female, the males will reach sexual maturity, developing large testicles as their other organs atrophy. This process allows for sperm to be in constant supply when the female produces an egg, so that the chimeric fish is able to have a greater number of offspring.

Germline Chimerism

Germline chimerism occurs when the germ cells (for example, sperm and egg cells) of an organism are not genetically identical to its own. It has recently been discovered that marmosets can carry the reproductive cells of their (fraternal) twin siblings, because of placental fusion during development. (Marmosets almost always give birth to fraternal twins.)

Human Chimeras

In 1953 a human chimera was reported in the British Medical Journal. A woman was found to have blood containing two different blood types. Apparently this resulted from cells from her twin brother living in her body. More recently, a study found that such blood group chimerism is not rare. Another report of a human chimera was published in 1998, where a male human had some partially-developed female organs due to chimerism. He was conceived by in vilro fertilization.

Chimeras in Research

In biological research, chimeras are artificially produced by selectively transplanting embryonic cells from one organism onto the embryo of another, and allowing the resultant blastocyst to develop. Chimeras are not hybrids, which form from the fusion of gametes from two species that form a single zygote with a combined genetic makeup, or Hybridomas which, as with hybrids, result from fusion of two species' cells into a single cell and artificial propagation of this cell in the laboratory. Essentially, in a chimera, each cell is from either of the parent species, whereas in a hybrid and hybridoma, each cell is derived from both parent species. "Chimera" is a broad term and is often applied to many different mechanisms of the mixing of cells from two different species.

As with cloning, the process of creating and implanting a chimera is imprecise, with the majority of embryos spontaneously terminating. Successes, however, have led to major advancements in the field of embryology, as creating chimeras of one species with different physical traits, such as colour, has allowed researchers to trace the differentiation of embryonic cells through the formation of organ systems in the adult individual.

A major milestone in chimera experimentation occurred in 1984, when a chimeric geep was produced by combining embryos from a goat and a sheep, and survived to adulthood. The creation of the "geep" revealed several complexities to chimera development. In implanting a goat embryo for gestation in a sheep, the sheep's immune system would reject the developing goat embryo, whereas a "geep" embryo, sharing markers of immunity with both sheep and goats, was able to survive implantation in either of its parent species.

In August 2003, researchers at the Shanghai Second Medical University in China reported that they had successfully fused human skin cells and dead rabbit eggs to create the first human chimeric embryos. The embryos were allowed to develop for several days in a laboratory setting, then destroyed to harvest the resulting stem cells. In 2007, scientists at the University of Nevada School of Medicine created a sheep whose blood contained 15% human cells and 85% sheep cells. The implications of increasingly realizable projects using human-animal hybrids for biopharmaceutical production, and potentially for producing cells or organs, have raised a host of ethical and safety issues.

Mouse Chimeras

Chimeric mice are important tools in biological research, as they allow the investigation of a variety of biological questions in an animal that has two distinct genetic pools within it. These include insights into such problems as the tissue specific requirements of a gene, cell lineage, and cell potential. The general methods for creating chimeric mice can be summarized either by injection or aggregation of embryonic cells from different origins.

The first chimeric mouse was made by Beatrice Mintz in the 1960s through the aggregation of eight cell stage embryos. Injection on the other hand was pioneered by Richard Gardner and Ralph Brinster who injected cells into blastocysts to create chimeric mice with germ lines fully derived from injected ES Cells. Mouse embryos both periimplantation and post-implantation can contribute to a chimera. It is post-implantation that ES cells from the inner cell mass of a blastocyst can contribute to all cell lineages of a mouse including the germ line. ES cells are also a useful tool in chimeras because genes can be mutated in them through the use of homologous recombination, thus allowing gene targeting. Since this discovery occurred in 1999, ES cells have become a key tool in the generation of specific chimeric mice.

Underlying Biology

The ability to make mouse chimeras comes from an understanding of early mouse development. Between the stages of fertilization of the egg and the implantation of a blastocyst into the uterus, different parts of the mouse embryo retain the ability to give rise to a variety of cell lineages. Once the embryo has reached the blastocyst stage, it is composed of several parts, mainly the trophectoderm, the inner cell mass, and the primitive endoderm. Each of these parts of the blastocyst gives rise to different parts of the embryo; the inner cell mass gives rise to the embryo proper, while the trophectoderm and primitive endoderm give rise to extra embryonic structures that support growth of the embryo. Two- to eight-cell-stage embryos are competent for making chimeras, since at these stages of development, the cells in the embryos are not yet committed to give rise to any particular cell lineage, and could give rise to the inner cell mass or the trophectoderm. In the case where two diploid eight-cell-stage embryos are used to make a chimera, chimersim can be later found in the epiblast, primitive, endoderm and trophectoderm of the mouse blastocyst.

It is possible to dissect the embryo at other stages so as to accordingly give rise to one lineage of cells from an embryo selectively and not the other. For example, subsets of blastomeres can be used to give rise to chimera with specified cell lineage from one embryo. The Inner Cell Mass of a diploid blastocyst for example can be used to make a chimera with another blastocyst of eight-cell diploid embryo; the cells taken from the inner cell mass will give rise to the primitive endoderm and to the epiblast in the chimera mouse. From this knowledge, ES cell contributions to chimeras have been developed. ES cells can be used in combination with eight-cell-and two-cell-stage embryos to make chimeras and exclusively give rise to the embryo proper. Embryos that are to be used in chimeras can further be genetically altered in order to specifically contribute to only one part of chimera. An example is the chimera built off of ES cells and tetraploid embryos, tetraploid embryos which are artificially made by electrofusion of two two-cell diploid embryos. The tetraploid embryo will exclusively give rise to the trophectoderm and primitive endoderm in the chimera

Methods of Production

There are a variety of combinations that can give rise to a successful chimera mouse and — according to the goal of the experiment — an appropriate cell and embryo combination can be picked; they are generally but not limited to diploid embryo and ES cells, diploid embryo and diploid embryo, ES cell and tetraploid embryo, diploid embryo and tetraploid embryo, ES cells and ES cells. The combination of embryonic stem cell and diploid embryo is a common technique used for the making of chimeric mice, since gene targeting can be done in the embryonic stem cell.

These kinds of chimeras can be made through either aggregation of stem cells and the diploid embryo or injection of the stem cells into the diploid embryo. If embryonic stem cells are to be used for gene targeting to make a chimera, the following procedure is common: a construct for homologous recombination for the gene targeted will be introduced into cultured mouse embryonic stem cells from the donor mouse, by way of electroporation; cells positive for the recombination event will have antibiotic resistance, provided by the insertion cassette used in the gene targeting; and be able to be positively selected for. ES cells with the correct targeted gene are then injected into a diploid host mouse blastocyst.

These injected blastocysts are then implanted into a pseudo pregnant female surrogate mouse which will bring the embryos to term and give birth to a mouse whose germline is derived from the donor mouse's ES cells. This same procedure can be achieved through aggregation of ES cells and diploid embryos, diploid embryos are cultured in aggregation plates in wells were single embryos can fit, to these wells ES cells are added the aggregates are cultured until a single embryo is formed and has progressed to the blastocyst stage, and can then be transferred to the surrogate mouse.

Legislation

The US and Western Europe have strict codes of ethics and regulations in place that expressly forbid certain subsets of experimentation using human cells, though there is a vast difference in the regulatory framework. In May 2008, a robust debate in the House of Commons of the United Kingdom on the ethics of creating chimeras with human stem cells led to the decision that embryos would be allowed to be made in laboratories, given that they would be destroyed within the first 14 days. No such foundation has been set for chimera research regulation in the US.

Reading Scientific Services (detecting GMOs)

Reading Scientific Services Ltd. (RSSL) is a British company providing scientific analysis, consultancy, product development and training to the food, healthcare, pharmaceutical, personal care, cosmetics and consumer goods industries. It has worked in areas such as the detection of contaminants including dioxins, benzene, melamine, unauthorised colours such as Sudan, undeclared genetically modified organisms, and foreign objects, with an emergency response service for urgent investigations. It also tests pharmaceutical packaging.

The company was formed in 1987 out of Cadbury Schweppes Research and Development. It is a wholly owned subsidiary of Cadbury. It has been inspected by regulatory authorities including the U.S. Food and Drug Administration (FDA), the Medicines and Healthcare products Regulatory Agency (MHRA) and the United Kingdom Accreditation Service (UKAS).

RSSL is based at two sites on the Whiteknights Campus of the University of Reading. The headquarters and main laboratories are based at the Reading Science Centre, whilst the training centre and clinical sciences are located at the Science & Technology Centre.

Smart Breeding

SMART breeding (SMART=Selection with Markers and Advanced Reproductive Technologies) or Precision breeding refers to a genetic engineering technique of reproducing a species members together to retain desirable traits and so produce a stronger hybrid. The technique was successfully used by Nachum Kedar, an Israeli scientist, who applied the technique using beefsteak tomatoes to produce a fruit that would ripen on the vine and remain firm in transit. Smart breeding has been advanced as an alternative to transgenics as a way to produce plants that are resistant to various environmental problems.

Sperm-mediated Gene Transfer

Sperm-mediated gene transfer, or SMGT, is a technique used to produce transgenic animals. Genetic material is introduced into sperm, which are used to fertilize eggs. The embryos are carried to term. The offspring may be transgenic, mosaic, or neither.

Synthetic Biology

Synthetic biology is a new area of biological research that combines science and engineering. It encompasses a variety of different approaches, methodologies, and disciplines with a variety of definitions. What they all have in common, however, is that they see synthetic biology as the design and construction of new biological functions and systems not found in nature.

History

The term "synthetic biology" has a history spanning the twentieth century. The first use was in Stéphane Leducs's publication of « Théorie physico-chimique de la vie et générations spontanées » (1910) and « La Biologie Synthétique » (1912). In 1974, the Polish geneticist Waclaw Szybalski used the term "synthetic biology", writing:

Let me now comment on the question "what next". Up to now we are working on the descriptive phase of molecular biology. ... But the real challenge will start when we enter the synthetic biology phase of research in our field. We will then devise new control elements and add these new modules to the existing genomes or build up wholly new genomes. This would be a field with the unlimited expansion potential and hardly any limitations to building "new better control circuits" and ..... finally other "synthetic" organisms, like a "new better mouse". ... I am not concerned that we will run out of exciting and novel ideas, ... in the synthetic biology, in general.

When in 1978 the Nobel Prize in Physiology or Medicine was awarded to Arber, Nathans and Smith for the discovery of restriction enzymes, Waclaw Szybalski wrote in an editorial comment in the journal Gene: The work on restriction nucleases not only permits us easily to construct recombinant DNA molecules and to analyze individual genes, but also has led us into the new era of synthetic biology where not only existing genes are described and analyzed but also new gene arrangements can be constructed and evaluated.

Perspectives

Biology

Biologists are interested in learning more about how natural living systems work. One simple, direct way to test our current understanding of a natural living system is to build an instance (or version) of the system in accordance with our current understanding of the system. Michael Elowitz's early work on the Repressilator is one good example of such work. Elowitz had a model for how gene expression should work inside living cells.

To test his model, he built a piece of DNA in accordance with his model, placed the DNA inside living cells, and watched what happened. Slight differences between observation and expectation highlight new science that may be well worth doing. Work of this sort often makes good use of mathematics to predict and study the dynamics of the biological system before experimentally constructing it. A wide variety of mathematical descriptions have been used with varying accuracy, including graph theory, Boolean networks, ordinary differential equations, stochastic differential equations, and Master equations (in order of increasing accuracy). Good examples include the work of Adam Arkin, Jim Collins and Alexander van Oudenaarden. See also the PBS Nova special on artificial life.

Chemistry

Biological systems are physical systems that are made up of chemicals. Around the turn of the 20th century, the science of chemistry went through a transition from studying natural chemicals to trying to design and build new chemicals. This transition led to the field of synthetic chemistry. In the same tradition, some aspects of synthetic biology can be viewed as an extension and application of synthetic chemistry to biology, and include work ranging from the creation of useful new biochemicals to studying the origins of life. Eric Kool's group at Stanford, the Foundation for Applied Molecular Evolution, Carlos Bustamante's group at Berkeley, Jack Szostak's group at Harvard, and David McMillen's group at University of Toronto are good examples of this tradition. Much of the improved economics and versatility of synthetic biology is driven by ongoing improvements in gene synthesis.

Engineering

Engineers view biology as a technology - the systems biotechnology or systems biological engineering . Synthetic Biology includes the broad redefinition and expansion of biotechnology, with the ultimate goals of being able to design and build engineered biological systems that process information, manipulate chemicals, fabricate materials and structures, produce energy, provide food, and maintain and enhance human health and our environment. A good example of these technologies include the work of Chris Voigt, who redesigned the Type III secretion system used by Salmonella typhimurium to secrete spider silk proteins, a strong elastic biomaterial, instead of its own natural infectious proteins. One aspect of Synthetic Biology which distinguishes it from conventional genetic engineering is a heavy emphasis on developing foundational technologies that make the engineering of biology easier and more reliable. Good examples of engineering in synthetic biology include the pioneering work of Tim Gardner and Jim Collins on an engineered genetic toggle switch, a riboregulator, the Registry of Standard Biological Parts, and the International Genetically Engineered Machine competition (iGEM).

Studies in synthetic biology can be subdivided into broad classifications according to the approach they take to the problem at hand: photocell design, biomolecular engineering, genome engineering, and biomolecular-design. The photocell approach includes projects to make self-replicating systems from entirely synthetic components. Biomolecular engineering includes approaches which aim to create a toolkit of functional units that can be introduced to present new orthogonal functions in living cells. Genome engineering includes approaches to construct synthetic chromosomes for whole or minimal organisms.

Biomolecular-design approach refers to the general idea of the de novo design and combination of biomolecular components. The task of each of these approaches is similar: To create a more synthetic entry at a higher level of complexity by manipulating a part of the proceeding level.

Re-writing

Re-writers are Synthetic Biologists who are interested in testing the idea that since natural biological systems are so complicated, we would be better off re-building the natural systems that we care about, from the ground up, in order to provide engineered surrogates that are easier to understand and interact with. Re-writers draw inspiration from refactoring, a process sometimes used to improve computer software. Drew Endy and his group have done some preliminary work on re-writing (e.g., Refactoring Bacteriophage T7). Oligonucleotides harvested from a photolithographic or inkjet manufactured DNA chip combined with DNA mismatch error-correction allows inexpensive large-scale changes of codons in genetic systems to improve gene expression or incorporate novel amino-acids (see George Church's and Anthony Forster's synthetic cell projects. As in the T7 example above, this favors a synthesis-from-scratch approach.

Key Enabling Technologies

There are several key enabling technologies that are critical to the growth of synthetic biology. The key concepts include standardization of biological parts and hierarchical abstraction to permit using those parts in increasingly complex synthetic systems. Achieving this is greatly aided by basic technologies of reading and writing of DNA (sequencing and fabrication), which are improving in price/performance exponentially (Kurzweil, 2001). Measurements under a variety of conditions are needed for accurate modeling and computer-aided-design (CAD).

DNA Sequencing

DNA sequencing is determining the order of the nucleotide bases in a molecule of DNA. Synthetic biologists make use of DNA sequencing in their work in several ways. First, large-scale genome sequencing efforts continue to provide a wealth of information on naturally occurring organisms. This information provides a rich substrate from which synthetic biologists can construct parts and devices. Second, synthetic biologists use sequencing to verify that they fabricated their engineered system as intended. Third, fast, cheap and reliable sequencing can also facilitate rapid detection and identification of synthetic systems and organisms.

Fabrication

A critical limitation in synthetic biology today is the time and effort expended during fabrication of engineered genetic sequences. To speed up the cycle of design, fabrication, testing and redesign, synthetic biology requires more rapid and reliable de novo DNA synthesis and assembly of fragments of DNA, in a process commonly referred to as gene synthesis.

In 2000, researchers at Washington University, mentioned synthesis of the 9.6 kbp Hepatitis C virus genome from chemically synthesized 60 to 80-mersIn 2002 researchers at SUNY Stony Brook succeeded in synthesizing the 7741 base poliovirus genome from its published sequence, producing the second synthetic genome. This took about two years of painstaking work. In 2003 the 5386 bp genome of the bacteriophage Phi X 174 was assembled in about two weeks. In 2006, the same team, at the J.

Craig Venter Institute, has constructed and patented a synthetic genome of a novel minimal bacterium, Mycoplasma laboratorium and is working on getting it functioning in a living cell.

In 2007 it was reported that several companies were offering the synthesis of genetic sequences up to 2000 bp long, for a price of about $1 per base pair and a turnaround time of less than two weeks. By September 2009, the price had dropped to less than $0.50 per base pair with some improvement in turn around time. Not only is the price judged lower than the cost of conventional cDNA cloning, the economics make it practical for researchers to design and purchase multiple variants of the same sequence to identify genes or proteins with optimized performance.

In 2010, Venter's group announced they had been able to assemble a complete genome of millions of base pairs, insert it into a cell, and cause that cell to start replicating.

Modeling

Models inform the design of engineered biological systems by allowing synthetic biologists to better predict system behavior prior to fabrication. Synthetic biology will benefit from better models of how biological molecules bind substrates and catalyze reactions, how DNA encodes the information needed to specify the cell and how multi-component integrated systems behave. Recently, multiscale models of gene regulatory networks have been developed that focus on synthetic biology applications. Simulations have been used that model all biomolecular interactions in transcription, translation, regulation, and induction of gene regulatory networks, guiding the design of synthetic systems.

Measurement

Precise and accurate quantitative measurements of biological systems are crucial to improving understanding of biology. Such measurements often help to elucidate how biological systems work and provide the basis for model construction and validation.

Differences between predicted and measured system behavior can identify gaps in understanding and explain why synthetic systems don't always behave as intended. Technologies which allow many parallel and time-dependent measurements will be especially useful in synthetic biology. Microscopy and flow cytometry are examples of useful measurement technologies.

Examples

Molecular cloning is a method used frequently by geneticists to obtain large quantities of a particular strand of DNA. It involves shaping a selected piece of DNA and inserting it into the DNA of a bacterium called a plasmid.

Once the alien DNA is inserted the bacteria is allowed to replicate thus replicating the DNA that it contains. After replication is completed the copies of foreign DNA are separated from the plasmid. In this sense the bacteria becomes a cyborg because a foreign element is introduced and interacts with the bacteria.

The Sleeping Beauty (58) transposon system is an example of an engineered enzyme for inserting precise DNA sequences into genomes of vertebrate animals. The SB transposon is a synthetic sequence that was created based on deriving a consensus sequence of extinct Tc1/mariner-type transposons that are found as evolutionary relics in the genomes of most, if not all, vertebrates. This enzyme took about a year to engineer and since its creation has been used for gene transfer, gene discovery, and gene therapy applications

Biosensor technology is another example of cyborg bacteria. One such sensor created in Oak Ridge National Laboratory and named "critter on a chip" used a coating of bioluminescent bacteria on a light sensitive computer chip to detect certain petroleum pollutants. When the bacteria sense the pollutant, it lights up and is then processed or amplified.

In Australia, biosensors have been created to detect viruses, bacteria, hormones, drugs, and DNA sequences. In the future scientists hope to create chips that can sense toxins such as environmental estrogens and warfare agents. Even more recently chemists at the University of Nebraska created a humidity gauge by using gold plated bacteria on a silicon chip. With a decrease in humidity there was an increase in the circuit flow. One unique feature that separates the chip from the bioluminescent ones is that that after it has been assimilated the bacteria no longer needs to be kept alive for the humidity gauge to work.

Nanotechnology also has made advances by using cyborgs.. Researchers at the École polytechnique de Montréal in Canada have attached a microscopic bead to swimming bacteria. Using a magnetic resonance imaging machine (MRI) the researchers have been able to use the magnetic properties of the bacteria to direct it to certain locations. The bead has no purpose at the moment but researchers hope store drugs or other viral fighting agents inside so that it may be released at the directed location.

Challenges

Opposition to Synthetic Biology

Opposition by civil society groups to Synthetic Biology has been led by the ETC Group who have called for a global moratorium on developments in the field and for no synthetic organisms to be released from the lab. In 2006 38 civil society organizations authored an open letter opposing voluntary regulation of the field and in 2008 ETC Group released the first critical report on the societal impacts of synthetic biology which they dubbed "Extreme Genetic Engineering".

Safety and Security

In addition to numerous scientific and technical challenges, synthetic biology raises questions for ethics, biosecurity, biosafety, involvement of stakeholders and intellectual property. To date, key stakeholders (especially in the US) have focused primarily on the biosecurity issues, especially the so-called dual-use challenge. For example, while the study of synthetic biology may lead to more efficient ways to produce medical treatments (e.g. against malaria), it may also lead to synthesis or redesign of harmful pathogens (e.g., smallpox) by malicious actors.

Proposals for licensing and monitoring the various phases of gene and genome synthesis began to appear in 2004. A 2007 study compared several policy options for governing the security risks associated with synthetic biology.

Other initiatives, such as OpenWetWare, diybio, biopunk, biohack, and possibly others, have attempted to integrate self-regulation in their proliferation of open source synthetic biology projects. However the distributed and diffuse nature of open-source biotechnology may make it more difficult to track, regulate, or mitigate potential biosafety and biosecurity concerns.

An initiative for self-regulation has been proposed by the International Association Synthetic Biology that suggests some specific measures to be implemented by the synthetic biology industry, especially DNA synthesis companies. Some scientists, however, argue for a more radical and forward looking approaches to improve safety and security issues.

They suggest to use not only physical containment as safety measures, but also trophic and semantic containment. Trophic containment includes for example the design of new and more robust forms of auxotrophy, while semantic containment means the design and construction of completely novel orthogonal life-forms.

Social and Ethical

Online discussion of "societal issues" took place at the SYNBIOSAFE forum on issues regarding ethics, safety, security, IPR, governance, and public perception (summary paper). On July 9–10, 2009, the National Academies' Committee of Science, Technology & Law convened a symposium on "Opportunities and Challenges in the Emerging Field of Synthetic Biology" (transcripts, audio, and presentations available).

Some efforts have been made to engage social issues "upstream" focus on the integral and mutually formative relations among scientific and other human practices. These approaches attempt to invent ongoing and regular forms of collaboration among synthetic biologists, ethicists, political analysts, funders, human scientists and civil society activists. These collaborations have consisted either of intensive, short term meetings, aimed at producing guidelines or regulations, or standing committees whose purpose is limited to protocol review or rule enforcement. Such work has proven valuable in identifying the ways in which synthetic biology intensifies already-known challenges in rDNA technologies. However, these forms are not suited to identifying new challenges as they emerge, and critics worry about uncritical complicity.

An example of efforts to develop ongoing collaboration is the "Human Practices" component of the Synthetic Biology Engineering Research Center in the US and the SYNBIOSAFE project in Europe, coordinated by IDC, that investigated the biosafety, biosecurity and ethical aspects of synthetic biology. A report from the Woodrow Wilson Center and the Hastings Center, a prestigious bioethics research institute, found that ethical concerns in synthetic biology have received scant attention.

In January 2009, the Alfred P. Sloan Foundation funded the Woodrow Wilson Center, the Hastings Center, and the J. Craig Venter Institute to examine the public perception, ethics, and policy implications of synthetic biology. Public perception and communication of synthetic biology is the main focus of COSY: Communicating Synthetic Biology, that showed that in the general public synthetic biology is not seen as too different from 'traditional' genetic engineering. To better communicate synthetic biology and its societal ramifications to a broader public, COSY and SYNBIOSAFE published a 38 min documentary film in October 2009 .

After a series of meetings in the fall of 2010, the Presidential Commission for the study of Bioethical Issues released a report, on December 16, to the President calling for enhanced Federal oversight in the emerging field of synthetic biology. The panel that facilitated the production of the report, composed of 13 scientists, ethicists, and public policy experts, said that the very newness of the science, which involves the design and construction of laboratory-made biological parts, gives regulators, ethicists and others time to identify problems early on and craft solutions that can harness the technology for the public good.

"We comprehensively reviewed the developing field of synthetic biology to understand both its potential rewards and risks," said Dr. Amy Gutmann, the Commission Chair and President of the University of Pennsylvania. "We considered an array of approaches to regulation—from allowing unfettered freedom with minimal oversight and another to prohibiting experiments until they can be ruled completely safe beyond a reasonable doubt. We chose a middle course to maximize public benefits while also safeguarding against risks." Dr. Gutmann said the Commission's approach recognizes the great potential of synthetic biology, including life saving medicines, and the generally distant risks posed by the field's current capacity. "Prudent vigilance suggests that federal oversight is needed and can be exercised in a way that is consistent with scientific progress," she said.

Oscillating Gene

In molecular biology, an oscillating gene or clock gene is a gene that is expressed in an oscillating pattern, often circadian. Clock (Circadian Locomotor Output Cycles Kaput) is a gene encoding a basic-helix-loop-helix-PAS transcription factor (CLOCK) that affects both the persistence and period of circadian rhythms. CLOCK functions as an essential activator of downstream elements in the pathway critical to the generation of circadian rhythms.

Discovery and Function

The Clock gene was first identified in 1994 by Dr. Joseph Takahashi and his colleagues. Takahashi used forward mutagenesis screening of mice treated with N-ethyl-N-nitrosourea to create and identify mutations in key genes that broadly affect circadian activity. The Clock mutants discovered through the screen displayed an abnormally long period of daily activity. This trait proved to be heritable, and mice bred to be homozygous for the mutation eventually lost all circadian rhythmicity after several days in constant darkness. This showed that intact Clock genes are necessary for normal mammalian circadian function. CLOCK protein has been found to play a central role as a transcription factor in the circadian pacemaker. In Drosophila, newly synthesized CLOCK (CLK) is hypophosphorylated in the cytoplasm before entering the nucleus.

Once in the nuclei, CLK is localized in nulear foci and is later redistributed homogeneously. CYCLE(CYC) (also known as dBMAL for the BMAL1 ortholog in mammals) dimerizes with CLK via their respective PAS domains. This dimer then recruits co-activator CREB-binding protein(CBP) and is further phosphorylated. Once phosphorylated, this CLK-CYC complex binds to the E-box elements of the promoters of period (per) and timeless (tim) via its bHLH domain, causing the stimulation of gene expression of per and tim.

A large molar excess of period (PER) and timeless (TIM) proteins causes formation of the PER-TIM heterodimer which prevents the CLK-CYC heterodimer from binding to the E-boxes of per and tim, essentially blocking per and tim transcription. CLK is hyperphosphorylated when double-time (DBT) kinase interacts with the CLK-CYC complex in a PER reliant manner, destabilizing both CLK and PER, leading to the degradation of both proteins. Hypophosphorylated CLK then accumulates, binds to the E-boxes of per and tim and activates their transcription once again. This cycle of post-translational phosphorylation suggest that temporal phosphorylation of CLK helps in the timing mechanism of the circadian clock.

A similar model is found in mice, in which BMAL1 dimerizes with CLOCK to activate per and cryptochrome(cry) transcription. PER and CRY proteins form a heterodimer which acts on the CLOCK-BMAL heterodimer to repress the transcription of per and cry.

CLOCK exhibits histone acetyl transferase (HAT) activity, which is enhanced by dimerization with BMAL1. Dr. Paolo Sassone-Corsi and colleagues demonstrated in vitro that CLOCK mediated HAT activity is necessary to rescue circadian rhythms in Clock mutants.

Mutants of Clock

Clock mutant organisms can either possess a null mutation or an antimorphic allele at the Clock locus that codes for an antagonist to the wild-type protein. The presence of an antimorphic protein downregulates the transcriptional products normally upregulated by Clock.

Drosophila

In Drosophila, a mutant form of Clock (Jrk) resulting from a premature stop codon that eliminates the activation domain of the CLOCK protein, causes dominant effects: half of the heterozygous flies with this mutant gene have an altered period while the other half become arrhythmic. Homozygous flies are all arrhythmic. Furthermore, these mutant flies express low levels of PER and TIM proteins, indicating that Clock functions as a positive element in the circadian loop.

Mice

The mouse homolog to the Jrk mutant is the ClockÄ19 mutant that possesses a deletion in exon 19 of the Clock gene. This dominant-negative mutation results in a defective CLOCK-BMAL dimer, which causes mice to have a decreased ability to activate per transcription. In constant darkness, ClockÄ19 mice heterozygous for the Clock mutant allele exhibit lengthened circadian periods, while ClockÄ19/Ä19 mice homozygous for the allele become arrhythmic. In both heterozygotes and homozygotes, this mutation also produces lengthened periods and arrhythmicity at the single-cell level.

Clock -/- null mutant mice, in which Clock has been knocked out, display completely normal circadian rhythms. The discovery of a null Clock mutant with a wild-type phenotype directly challenged the widely accepted premise that Clock is necessary for normal circadian function. Furthermore, it suggested that the CLOCK-BMAL1 dimer need not exist to modulate other elements of the circadian pathway. Neuronal PAS domain containing protein 2 (NPAS2, a CLOCK analog) can substitute for CLOCK in these Clock-null mice.

Clock's Role in Other Feedback Loops

The CLOCK-BMAL dimer is involved in regulation of other genes and feedback loops. An enzyme SIRT1 has also binds to the CLOCK-BMAL complex and acts to suppress its activity, perhaps by deacetylation of Bmal1 and surrounding histones. However, SIRT1's role is still controversial and it may also have a role in deacetylating PER protein, targeting it for degradation.

The CLOCK-BMAL dimer acts as a positive limb of a feedback loop. The binding of CLOCK-BMAL to an E-box promoter element activates transcription of clock genes such as per1, 2, and 3 and tim in mice. It has been shown in mice that CLOCK-BMAL also activates the Nicotinamide phosphoribosyltransferase gene (also called Nampt), part of a separate feedback loop. This feedback loops creates a metabolic oscillator. The CLOCK-BMAL dimer activates transcription of the Nampt gene, which codes for the NAMPT protein. NAMPT is part of a series of enzymatic reactions that covert niacin (also called nicotinamide) to NAD. SIRT1, which requires NAD for its enzymatic activity, then uses increased NAD levels to deacetylate BMAL1, suppressing it.

This suppression results in less transcription of the NAMPT, less NAMPT protein, less NAD made, and therefore less SIRT1 and less suppression of the

CLOCK-BMAL dimer. This dimer can again positively activate the Nampt gene transcription and the cycle continues, creating another oscillatory loop involving CLOCK-BMAL as positive elements. The key role that Clock plays in metabolic and circadian loops highlights the close relationship between metabolism and circadian clocks.

Other Functions of Clock

In humans, a polymorphism in Clock, rs6832769, may be related to the personality trait agreeableness. Another single nucleotide polymorphism (SNP) in Clock, 3111C, has been associated with diurnal preference. This SNP is also associated with increased insomnia, difficulty losing weight, and recurrence of major depressive episodes in patients with bipolar disorder. In mice, Clock has been implicated in sleep disorders, metabolism, pregnancy, and mood disorders. Clock mutant mice sleep less than normal mice each day. The mice also display altered levels of plasma glucose and rhythms in food intake. These mutants develop metabolic syndrome symptoms over time.

Furthermore, Clock mutants demonstrate disrupted estrous cycles and increased rates of full-term pregnancy failure. Mutant Clock has also been linked to bipolar disorder-like symptoms in mice, including mania and euphoria. Clock mutant mice also exhibit increased excitability of dopamine neurons in reward centers of the brain. These results have led Dr. Colleen McClung to propose using Clock mutant mice as a model for human mood and behavior disorders.

The CLOCK-BMAL dimer has also been shown to activate reverse-erb receptor alpha (Rev-ErbA alpha) and retinoic acid orphan receptor alpha (ROR-alpha). REV-ERBá and RORá regulate Bmal by binding to retinoic acid-related orphan receptor response elements (ROREs) in its promoter.

In 2010, a Yale University team led by Dr. Yong Zhu found that variations in the epigenetics of the Clock gene may lead to an increased risk of breast cancer. It was found that in women with breast cancer, there was significantly less methylation of the Clock promoter region. It was also noted that this effect was greater in women with estrogen and progesterone receptor-negative tumors.

The CLOCK gene may also be a target for somatic mutations in microsatellite unstable colorectal cancers. In a study done in 2010 by researchers in the University of Helsinki, 53% of putative novel microsatellite instability target genes responsible for colorectal cancer contained CLOCK mutations.

Period (Gene)

Period (per) is a gene located on the X chromosome of Drosophila melanogaster. Oscillations in levels of both per transcript and its corresponding protein PER have a period of approximately 24 hours and together play a central role in the molecular mechanism of the Drosophila biological clock driving circadian rhythms in eclosion and locomotor activity. Mutations in the per gene can shorten (per), lengthen (per), and even abolish (per) the period of the circadian rhythm.

Discovery

The period gene and three mutants (per, per, and per) were isolated in an EMS mutagenesis screen by Ronald Konopka and Seymour Benzer in 1971. The per, per, and per mutations were found to complement each other, so it was concluded that the three phenotypes were due to mutations in the same gene. The discovery of mutants that altered the period of circadian rhythms in eclosion and locomotor activity (per and per) indicated the role of the per gene in the clock itself and not an output pathway. The period gene was first sequenced in 1984 by Michael Rosbash and colleagues. In 1998, it was discovered that per produces two transcripts (differing only by the alternative splicing of a single untranslated intron) which both encode the PER protein.

Function

Circadian Clock

In Drosophila, per mRNA levels oscillate with a period of approximately 24 hours, peaking during the early subjective night. The per product PER also oscillates with a nearly 24 hour period, peaking about six hours after per mRNA levels during the middle subjective night.

When PER levels increase, the inhibition of per transcription increases, lowering the protein levels. However, because PER protein cannot directly bind to DNA, it does not directly influence its own transcription; alternatively, it inhibits its own activators. After PER is produced from per mRNA, it dimerizes with Timeless (TIM) and the complex goes into the nucleus and inhibits the transcription factors of per and tim, the CLOCK/CYCLE heterodimer.

This CLOCK/CYCLE complex acts as a transcriptional activator for per and tim by binding to specific enhancers (called E-boxes) of their promoters. Therefore, inhibition of CLK/CYC lowers per and tim mRNA levels, which in turn lower the levels of PER and TIM. Now, cryptochrome (CRY) is a light sensitive protein which inhibits TIM in the presence of light. When TIM is not complexed with PER, another protein, doubletime, or DBT, phosphorylates PER, targeting it for degradation.

In mammals, an analogous transcription-translation negative feedback loop is observed. Translated from the three mammalian homologs of drosophila-per, the three PER proteins (PER1, PER2, and PER3) dimerize via their PAS domains with two cryptochrome proteins (CRY1 and CRY2) to form a negative element of the clock. This PER/CRY complex moves into the nucleus upon phosphorylation by CK1-epsilon (casein kinase 1 epsilon) and inhibits the CLK/BMAL1 heterodimer, the transcription factor that is bound to the E-boxs of the three per and two cry promoters by basic helix-loop-helix (BHLH) DNA-binding domains.

The mammalian period 1 and period 2 genes play key roles in photoentrainment of the circadian clock to light pulses. This was first seen in 1999 when Akiyama et al. showed that mPer1 is necessary for phase shifts induced by light or glutamate release. Two years later, Albrecht et al. found genetic evidence to support this result when they discovered that mPer1 mutants are not able to advance the clock in response to a late night light pulse (ZT22) and that mPer2 mutants are not able to delay the clock in response to an early night light pulse (ZT14). Thus, mPer1 and mPer2 are necessary for the daily resetting of the circadian clock to normal environmental light cues. per has also been implicated in the regulation of several output processes of the biological clock, including mating activity and oxidative stress response, through per mutation and knockout experiments.

Drosphila melanogaster has naturally occurring variation in Thr-Gly repeats, occurring along a latitude cline. Flies with 17 Thr-Gly repeats are found more commonly in Southern Europe and 20 Thr-Gly repeats are found more commonly in Northern Europe.

Non-circadian

In addition to its circadian functions, per has also been implicated in a variety of other non-circadian processes.

The mammalian period 2 gene plays a key role in tumor growth in mice; mice with an mPer2 knockout show a significant increase in tumor development and a significant decrease in apoptosis. This is thought to be caused by mPer2 circadian deregulation of common tumor suppression and cell cycle regulation genes, such as Cyclin D1, Cyclin A, Mdm-2, and Gadd45á, as well as the transcription factor c-myc, which is directly controlled by circadian regulators through E box-mediated reactions. In addition, mPer2 knockout mice show increased sensitivity to gamma radiation and tumor development, further implicating mPer2 in cancer development through its regulation of DNA damage-responsive pathways. Thus, circadian control of clock controlled genes that function in cell growth control and DNA damage response may affect the development of cancer in vivo.

It has been shown to be necessary and sufficient for long-term memory (LTM) formation in Drosophila melanogaster. per mutants show deficiencies in LTM formation that can be rescued with the insertion of a per transgene and enhanced with overexpression of the per gene. This response is absent in mutations of other clock genes (timeless, dClock, and cycle). Research suggests that synaptic transmission through per-expressing cells is necessary for LTM retrieval.

It has also been shown to extend the lifespan of the fruit fly, suggesting a role in aging. This result, however, is still controversial, as the experiments have not been successfully repeated by another research group.

In mice it has been shown that there is a link between per2 and preferred alcohol intake. Alcohol consumption has also been linked to shortening the free running period. The effect of alcoholism on per1 and per2 genes have also linked to the depression associated with alcohol as well as an individual's disposition to relapse into alcoholism.

2

Molecular Biotechnology

Molecular biotechnology is an exciting revolutionary scientific discipline that is based on the ability of a researcher to transfer specific units of genetic information from one organism to another. The objective of recombinant DNA technology is often to produce a useful product or a commercial process. In early 1970, traditional biotechnology was not well known as a scientific discipline. Research in this area was carried out in chemical engineering departments.

The term "biotechnology" was created in 1919 by a Hungarian engineer, Karl Ereky. According to Ereky, "biotechnology involves all works carried out with the aid of living things." More formally, biotechnology may be defined as "the application of scientific and engineering principles to the processing of material by biological agents to provide goods and services." In 1961, a Swedish microbiologist Carl Gordon Heden redefined biotechnology as "the industrial production of goods and services by processing using biological organisms, and it is firmly ground on expertise in microbiology, biochemistry and chemical engineering." However the nature of biotechnology was changed forever by the development of recombinant DNA technology. With these techniques, the optimization of any biotechnological process was achieved more directly. Genetic engineering provided a means to create, rather than merely isolate highly productive strains.

Microorganisms and eukaryotic cells could be used as "biological factories" for the production of insulin, interferon, growth hormone, viral antigens and a variety of other proteins. Recombinant DNA technology could also be used to facilitate the biological production of large amounts of useful lower molecular weight compounds and macromolecules that occur naturally in minute quantities. This, technology facilitated the development of radically new medical therapies and diagnostic systems. Thus, the union of recombinant DNA technology with biotechnology created a vibrant, highly competitive field of study that has been called "Molecular Biotechnology." Molecular biotechnology ought to contribute unprecedented benefits to humanity.

1. It should provide opportunities to accurately diagnose and prevent or cure a wide range of infectious and genetic diseases.
2. Significant increase in crop yield may be obtained by generating disease-pathogen- or herbicide-resistant varieties.
3. Microorganisms that will produce chemicals, antibiotics, polymers, amino acids, enzymes, etc. can be developed.
4. Livestock and other animals that have enhanced genetically determined attributes can be developed. Molecular biotechnology with much fuss and fanfare has become a comprehensive scientific venture, both commercially and academically, in a remarkably short time.
5. A number of new scientific and business publications are devoted to molecular biotechnology. Both graduate and undergraduate programmes and courses have been created at many universities throughout the world.

Although it is exciting and important to emphasize the positive aspects of new advances, there are also social concerns and consequences that must be addressed. Because molecular biotechnology is so broadly based, its potential impact on society must be considered.

Many issues have been considered and discussed extensively by government commissions, and debated by individuals in popular and academic publications. There has been an active and extensive participation by both scientists and the general public in deciding how molecular biotechnology should proceed, although some controversies still remain.

Finally, molecular biotechnology is the third scientific revolution after the industrial and computer revolution, which is going to change the lives and future of humankind on this earth. The ability to manipulate genetic material to achieve specified outcome in living organisms promises major changes in many aspects of modem life

The Technology that may Impact Our Life in Future!

"Distinct from traditional biotechnology, the modern methods require knowledge of and manipulation of genes, the functional units of inheritance, and the discipline that is concerned with the manipulation of genes for the purpose of producing useful goods and services using living organisms is known as molecular biotechnology. " — *Molecular Biotechnology*, 4th Edition, American Society for Microbiology

The Many Definitions of Molecular Biotechnology

- Molecular Biotechnology is a new research field that emerged as an outcome of recombinant DNA technology and traditional industrial microbiology.
- Molecular Biotechnology is an exciting field in science that is based on the ability of researchers to transfer specific units of genetic information from one organism to another, which relies on the technique of recombinant DNA technology (genetic engineering).
- Molecular Biotechnology is the use of laboratory techniques to study and find ways to modify nucleic acids and proteins for its applications in the areas such as medicine, animal health, agriculture, environment etc.
- Molecular Biotechnology is the result of fusion of areas of research, such as molecular biology, microbiology, biochemistry, immunology, genetics, and cell biology. It is an exciting field fuelled by the ability to transfer genetic information between organisms with the goal of understanding important biological processes or creating a useful product which benefits mankind.

DNA

Deoxyribonucleic acid or DNA, is a nucleic acid that contains the genetic instructions used in the development and functioning of all known living organisms (with the exception of RNA viruses).

The main role of DNA molecules is the long-term storage of information. DNA is often compared to a set of blueprints, like a recipe or a code, since it contains the instructions needed to construct other components of cells, such as proteins and RNA molecules. The DNA segments that carry this genetic information are called genes, but other DNA sequences have structural purposes, or are involved in regulating the use of this genetic information.

DNA consists of two long polymers of simple units called nucleotides, with backbones made of sugars and phosphate groups joined by ester bonds. These two strands run in opposite directions to each other and are therefore anti-parallel. Attached to each sugar is one of four types of molecules called bases. It is the sequence of these four bases along the backbone that encodes information. This information is read using the genetic code, which specifies the sequence of the amino acids within proteins. The code is read by copying stretches of DNA into the related nucleic acid RNA, in a process called transcription. Within cells, DNA is organized into long structures called chromosomes. These chromosomes are duplicated before cells divide, in a process called DNA replication. Eukaryotic organisms (animals, plants, fungi, and protists) store most of their DNA inside the cell nucleus and some of their DNA in organelles, such as mitochondria or chloroplasts.

In contrast, prokaryotes (bacteria and archaea) store their DNA only in the cytoplasm. Within the chromosomes, chromatin proteins such as histones compact and organize DNA. These compact structures guide the interactions between DNA and other proteins, helping control which parts of the DNA are transcribed.

Properties

DNA is a long polymer made from repeating units called nucleotides. As first discovered by James D. Watson and Francis Crick, the structure of DNA of all species comprises two helical chains each coiled round the same axis, and each with a pitch of 34 Angstroms (3.4 nanometres) and a radius of 10 Angstroms (1.0 nanometres). According to another study, when measured in a particular solution, the DNA chain measured 22 to 26 Angstroms wide (2.2 to 2.6 nanometres), and one nucleotide unit measured 3.3 Å (0.33 nm) long. Although each individual repeating unit is very small, DNA polymers can be very large molecules containing millions of nucleotides. For instance, the largest human chromosome, chromosome number 1, is approximately 220 million base pairs long.

In living organisms, DNA does not usually exist as a single molecule, but instead as a pair of molecules that are held tightly together. These two long strands entwine like vines, in the shape of a double helix. The nucleotide repeats contain both the segment of the backbone of the molecule, which holds the chain together, and a base, which interacts with the other DNA strand in the helix. A base linked to a sugar is called a nucleoside and a base linked to a sugar and one or more phosphate groups is called a nucleotide. If multiple nucleotides are linked together, as in DNA, this polymer is called a polynucleotide.

The backbone of the DNA strand is made from alternating phosphate and sugar residues. The sugar in DNA is 2-deoxyribose, which is a pentose (five-carbon) sugar. The sugars are joined together by phosphate groups that form phosphodiester bonds between the third and fifth carbon atoms of adjacent sugar rings. These asymmetric bonds mean a strand of DNA has a direction. In a double helix the direction of the nucleotides in one strand is opposite to their direction in the other strand: the strands are *antiparallel*. The asymmetric ends of DNA strands are called the 52 (*five prime*) and 32 (*three prime*) ends, with the 5' end having a terminal phosphate group and the 3' end a terminal hydroxyl group. One major difference between DNA and RNA is the sugar, with the 2-deoxyribose in DNA being replaced by the alternative pentose sugar ribose in RNA.

The DNA double helix is stabilized by hydrogen bonds between the bases attached to the two strands. The four bases found in DNA are adenine (abbreviated A), cytosine (C), guanine (G) and thymine (T). These four bases are attached to the sugar/phosphate to form the complete nucleotide, as shown for adenosine monophosphate.

These bases are classified into two types; adenine and guanine are fused five- and six-membered heterocyclic compounds called purines, while cytosine and thymine are six-membered rings called pyrimidines. A fifth pyrimidine base, called uracil (U), usually takes the place of thymine in RNA and differs from thymine by lacking a methyl group on its ring. Uracil is not usually found in DNA, occurring only as a breakdown product of cytosine. In addition to RNA and DNA, a large number of artificial nucleic acid analogues have also been created to study the proprieties of nucleic acids, or for use in biotechnology.

Grooves

Twin helical strands form the DNA backbone. Another double helix may be found by tracing the spaces, or grooves, between the strands. These voids are adjacent to the base pairs and may provide a binding site. As the strands are not directly opposite each other, the grooves are unequally sized. One groove, the major groove, is 22 Å wide and the other, the minor groove, is 12 Å wide. The narrowness of the minor groove means that the edges of the bases are more accessible in the major groove. As a result, proteins like transcription factors that can bind to specific sequences in double-stranded DNA usually make contacts to the sides of the bases exposed in the major groove. This situation varies in unusual conformations of DNA within the cell, but the major and minor grooves are always named to reflect the differences in size that would be seen if the DNA is twisted back into the ordinary B form.

Base Pair

In molecular biology, two nucleotides on opposite complementary DNA or RNA strands that are connected via hydrogen bonds are called a base pair (often abbreviated bp). In the canonical Watson-Crick DNA base pairing, adenine (A) forms a base pair with thymine (T) and guanine (G) forms a base pair with cytosine (C). In RNA, thymine is replaced by uracil (U). Alternate hydrogen bonding patterns, such as the wobble base pair and Hoogsteen base pair, also occur—particularly in RNA—giving rise to complex and functional tertiary structures. Importantly, pairing is the mechanism by which codons on messenger RNA molecules are recognized by anticodons on transfer RNA during protein translation. Some DNA- or RNA-binding enzymes can recognize specific base pairing patterns that identify particular regulatory regions of genes.

The size of an individual gene or an organism's entire genome is often measured in base pairs because DNA is usually double-stranded. Hence, the number of total base pairs is equal to the number of nucleotides in one of the strands (with the exception of non-coding single-stranded regions of telomeres). The haploid human genome (23 chromosomes) is estimated to be about 3 billion base pairs long and to contain 20,000–25,000 distinct genes. A kilobase (kb) is a unit of measurement in molecular biology equal to 1000 base pairs of DNA or RNA.

Hydrogen Bonding and Stability

Hydrogen bonding is the chemical interaction that underlies the base-pairing rules described above. Appropriate geometrical correspondence of hydrogen bond donors and acceptors allows only the "right" pairs to form stably. DNA with high GC-content is more stable than DNA with low GC-content, but contrary to popular belief, the hydrogen bonds do not stabilize the DNA significantly and stabilization is mainly due to stacking interactions.

The larger nucleobases, adenine and guanine, are members of a class of doubly-ringed chemical structures called purines; the smaller nucleobases, cytosine and thymine (and uracil), are members of a class of singly-ringed chemical structures called pyrimidines. Purines are only complementary with pyrimidines: pyrimidine-pyrimidine pairings are energetically unfavourable because the molecules are too far apart for hydrogen bonding to be established; purine-purine pairings are energetically unfavourable because the molecules are too close, leading to overlap repulsion. The only other possible pairings are GT and AC; these pairings are mismatches because the pattern of hydrogen donors and acceptors do not correspond. The GU pairing, with two hydrogen bonds, does occur fairly often in RNA.

Paired DNA and RNA molecules are comparatively stable at room temperature but the two nucleotide strands will separate above a melting point that is determined by the length of the molecules, the extent of mispairing (if any), and the GC content. Higher GC content results in higher melting temperatures; it is therefore unsurprising that the genomes of extremophile organisms such as *Thermus thermophilus* are particularly GC-rich. Conversely, regions of a genome that need to separate frequently—for example, the promoter regions for often-transcribed genes—are comparatively GC-poor. GC content and melting temperature must also be taken into account when designing primers for PCR reactions.

Base Stacking

Base stacking interactions in DNA and RNA are due to dispersion attraction, short-range exchange repulsion, and electrostatic interactions which also contribute to stability . Again GC stacking interactions with adjacent bases tend to be more favourable. (Note, though, that a GC stacking interaction with the next base pair is geometrically different from a CG interaction.) Base stacking effects are especially important in the secondary structure and tertiary structure of RNA; for example, RNA stem-loop structures are stabilized by base stacking in the loop region.

Base Analogs and Intercalators

Chemical analogs of nucleotides can take the place of proper nucleotides and establish non-canonical base-pairing, leading to errors (mostly point mutations) in DNA replication and DNA transcription. One common mutagenic base analog is 5-bromouracil, which resembles thymine but can base-pair to guanine in its enol form.

Other chemicals, known as DNA intercalators, fit into the gap between adjacent bases on a single strand and induce frameshift mutations by "masquerading" as a base, causing the DNA replication machinery to skip or insert additional nucleotides at the intercalated site. Most intercalators are large polyaromatic compounds and are known or suspected carcinogens. Examples include ethidium bromide and acridine.

Examples

The following DNA sequences illustrate pair double-stranded patterns. By convention, the top strand is written from the 5' end to the 3' end; thus the bottom strand is written 3' to 5'.

A base-paired DNA sequence:

ATCGATTGAGCTCTAGCG

TAGCTAACTCGAGATCGC

The corresponding RNA sequence, in which uracil is substituted for thymine where uracil takes its place in the RNA strand:

AUCGAUUGAGCUCUAGCG

UAGCUAACUCGAGAUCGC

Length Measurements

The following abbreviations are commonly used to describe the length of a D/RNA molecule:

- bp = base pair(s)—one bp corresponds to circa 3.4 Å of length along the strand
- kb (= kbp) = kilo base pairs = 1,000 bp
- Mb = mega base pairs = 1,000,000 bp
- Gb = giga base pairs = 1,000,000,000 bp.

In case of single stranded DNA/RNA units of nucleotides are used, abbreviated nt (or knt, Mnt, Gnt), as they are not paired. For distinction between units of computer storage and bases kbp, Mbp, Gbp etc. may be used for basepairs. The length of 16S rDNA for bacteria is 1542 base-pair in length.

The Centimorgan is also often used to imply distance along a chromosome, but the number of base-pairs it corresponds to varies widely. In the Human genome, it is about 1 million base pairs.

Sense (Molecular Biology)

Sense, when applied in a molecular biology context, is a general concept used to compare the polarity of nucleic acid molecules, such as DNA or RNA, to other nucleic acid molecules. Depending on the context within molecular biology, sense may have slightly different meanings.

DNA Sense

Molecular biologists call a single strand of DNA sense (or positive (+) sense) if an RNA version of the same sequence is translated or translatable into protein.

Its complementary strand is called antisense (or negative (-) sense). Sometimes the phrase coding strand is encountered; however, protein coding and non-coding RNA's can be transcribed similarly from both strands, in some cases being transcribed in both directions from a common promoter region, or being transcribed from within introns, on both strands.

Antisense DNA

DNA normally has two strands, i.e., the sense strand and the antisense strand. In double-stranded DNA, only one strand codes for the RNA that is translated into protein. This DNA strand is referred to as the antisense strand. The strand that does not code for RNA is called the sense strand because it has a similar sequence to the messenger RNA (mRNA). Both the sense DNA strand and the mRNA transcript are complementary to the template DNA strand. Note that the DNA strands called "sense" and "antisense" are sometimes switched in older textbooks.

Antisense molecules interact with complementary strands of nucleic acids, modifying expression of genes.

Example with Double Stranded DNA :

DNA strand 1: antisense strand (copied to)→ RNA strand (sense).

DNA strand 2: sense strand.

Some regions within a double strand of DNA code for genes, which are usually instructions specifying the order of amino acids in a protein along with regulatory sequences, splicing sites, noncoding introns, and other complicating details. For a cell to use this information, one strand of the DNA serves as a template for the synthesis of a complementary strand of RNA. The template DNA strand is called the transcribed strand with antisense sequence and the mRNA transcript is said to be sense sequence (the complement of antisense). Because the DNA is double-stranded, the strand complementary to the antisense sequence is called non-transcribed strand and has the same sense sequence as the mRNA transcript (though T bases in DNA are substituted with U bases in RNA).

A note on the confusion between "sense" and "antisense" strands: The strand names actually depend on which direction you are writing the sequence that contains the information for proteins (the "sense" information), not on which strand is on the top or bottom (that is arbitrary).

The only real biological information that is important for labeling strands is the location of the 5' phosphate group and the 3' hydroxyl group because these ends determine the direction of transcription and translation. A sequence 5' CGCTAT 3' is equivalent to a sequence written 3' TATCGC 5' as long as the 5' and 3' ends are noted.

If the ends are not labelled, convention is to assume that the sequence is written in the 5' to 3' direction. Good rule of thumb for figuring out the "sense" strand: Look for the start codon ATG (AUG in mRNA). In the table example, the sense mRNA has the AUG codon at the end (remember that translation proceeds in the 5' to 3' direction).

Ambisense

A single-stranded genome that contains both positive-sense and negative-sense is said to be ambisense. Bunya viruses have 3 single-stranded RNA (ssRNA) fragments containing both positive-sense and negative-sense sections; arenaviruses are also ssRNA viruses with an ambisense genome, as they have 2 fragments that are mainly negative-sense except for part of the 5' ends of the large and small segments of their genome.

Antisense RNA

Antisense RNA is a single-stranded RNA that is complementary to a messenger RNA (mRNA) strand transcribed within a cell. Antisense RNA may be introduced into a cell to inhibit translation of a complementary mRNA by base pairing to it and physically obstructing the translation machinery. This effect is therefore stoichiometric. An example of naturally occurring mRNA antisense mechanism is the hok/sok system of the E.coli R1 plasmid.

Antisense RNA has long been thought of as a promising technique for disease therapy; the only such case to have reached the market is the drug fomivirsen. One commentator has characterized antisense RNA as one of "dozens of technologies that are gorgeous in concept, but exasperating in [commercialization]". Generally, antisense RNA still lack effective design, biological activity, and efficient route of administration.

Historically, the effects of antisense RNA have often been confused with the effects of RNA interference (RNAi), a related process in which *double-stranded* RNA fragments called small interfering RNAs trigger catalytically mediated gene silencing, most typically by targeting the RNA-induced silencing complex (RISC) to bind to and degrade the mRNA. Attempts to genetically engineer transgenic plants to express antisense RNA instead activate the RNAi pathway, although the processes result in differing magnitudes of the same downstream effect, gene silencing. Well-known examples include the Flavr Savr tomato and two cultivars of ringspot-resistant papaya.

Transcription of longer cis-antisense transcripts is a common phenomenon in the mammalian transcriptome. Although the function of some cases have been described, such as the Zeb2/Sip1 antisense RNA, no general function has been elucidated. In the case of Zeb2/Sip1, the antisense noncoding RNA is opposite the 5' splice site of an intron in the 5'UTR of the Zeb2 mRNA. Expression of the antisense ncRNA prevents splicing of an intron that contains a ribosome entry site necessary for efficient expression of the Zeb2 protein. Transcription of long antisense ncRNAs is often concordant with the associated protein-coding gene, but more detailed studies have revealed that the relative expression patterns of the mRNA and antisense ncRNA are complex.

RNA Sense in Viruses

In virology, the genome of an RNA virus can be said to be either positive-sense, also known as a "plus-strand", or negative-sense, also known as a "minus-strand". In most cases, the terms *sense* and *strand* are used interchangeably, making such terms as *positive-strand* equivalent to *positive-sense*, and *plus-strand* equivalent to *plus-sense*. Whether a virus genome is positive-sense or negative-sense can be used as a basis for classifying viruses.

Positive-sense

Positive-sense (5' to 3') viral RNA signifies that a particular viral RNA sequence may be directly translated into the desired viral proteins. Therefore, in positive-sense RNA viruses, the viral RNA genome can be considered viral mRNA, and can be immediately translated by the host cell. Unlike negative-sense RNA, positive-sense RNA is of the same sense as mRNA. Some viruses (e.g., Coronaviridae) have positive-sense genomes that can act as mRNA and be used directly to synthesise proteins without the help of a complementary RNA intermediate. Because of this, these viruses do not need to have an RNA polymerase packaged into the virion.

Negative-sense

Negative-sense (3' to 5') viral RNA is complementary to the viral mRNA and thus must be converted to positive-sense RNA by an RNA polymerase prior to translation. Negative-sense RNA (like DNA) has a nucleotide sequence complementary to the mRNA that it encodes. Like DNA, this RNA cannot be translated into protein directly. Instead, it must first be transcribed into a positive-sense RNA that acts as an mRNA. Some viruses (Influenza, for example) have negative-sense genomes and so must carry an RNA polymerase inside the virion.

Enzymes

Enzyme-dependent antisense includes forms dependent on RNase H activity to degrade target mRNA, including single-stranded DNA, RNA, and phosphorothioate antisense. The R1 plasmid hok/sok system is an example of mRNA antisense regulation process, through enzymatic degradation of the resulting RNA duplex. Double stranded RNA acts as enzyme-dependent antisense through the RNAi/siRNA pathway, involving target mRNA recognition through sense-antisense strand pairing followed by target mRNA degradation by the RNA-induced silencing complex (RISC).

Steric blocking antisense (RNase-H-independent antisense) interferes with gene expression or other mRNA-dependent cellular processes by binding to a target sequence of mRNA and getting in the way of other processes. Steric blocking antisense includes 2'-O alkyl (usually in chimeras with RNase-H dependent antisense), peptide nucleic acid (PNA), locked nucleic acid (LNA), and Morpholino antisense.

DNA Supercoil

In a "relaxed" double-helical segment of B-DNA, the two strands twist around the helical axis once every 10.4-10.5 base pairs of sequence. Adding or subtracting twists, as some enzymes can do, imposes strain. If a DNA segment under twist strain were to be closed into a circle by joining its two ends and then it is allowed to move freely, the circular DNA would contort into a new shape. Such a contortion is a supercoil. For each additional helical twist being accommodated, the lobes will show one more rotation about their axis.

The noun form "supercoil" is rarely used in the context of DNA topology. Instead, global contortions of a circular DNA, such as the rotation, are referred to as *writhe*. "Supercoiling" is an abstract mathematical property representing the sum of twist and writhe. The twist is the number of helical turns in the DNA and the writhe is the number of times the double helix crosses over on itself (these are the supercoils).

Extra helical twists are positive and lead to positive supercoiling, while subtractive twisting causes negative supercoiling. Many topoisomerase enzymes sense supercoiling and either generate or dissipate it as they change DNA topology. DNA of most organisms is negatively supercoiled.

In part because chromosomes may be very large, segments in the middle may act as if their ends are anchored. As a result, they may be unable to distribute excess twist to the rest of the chromosome or to absorb twist to recover from underwinding—the segments may become *supercoiled*, in other words. In response to supercoiling, they will assume an amount of writhe, just as if their ends were joined.

Supercoiled DNA forms two structures; a plectoneme or a toroid, or a combination of both. A negatively supercoiled DNA molecule will produce either a one-start left-handed helix, the toroid, or a two-start right-handed helix with terminal loops, the plectoneme. Plectonemes are typically more common in nature, and this is the shape most bacterial plasmids will take.

For larger molecules it is common for hybrid structures to form - a loop on a toroid can extend into a plectoneme. If all the loops on a toroid extend then it becomes a branch point in the plectonemic structure.

Occurrence of DNA Supercoiling

DNA supercoiling is important for DNA packaging within all cells. Because the length of DNA can be thousands of times that of a cell, packaging this genetic material into the cell or nucleus (in eukaryotes) is a difficult feat. Supercoiling of DNA reduces the space and allows for a lot more DNA to be packaged. In prokaryotes, plectonemic supercoils are predominant, because of the circular chromosome and relatively small amount of genetic material. In eukaryotes, DNA supercoiling exists on many levels of both plectonemic and solenoidal supercoils, with the solenoidal supercoiling proving most effective in compacting the DNA. Solenoidal supercoiling is achieved with histones to form a 10nm fibre. This fibre is further coiled into a 30nm fibre, and further coiled upon itself numerous times more.

DNA packaging is greatly increased during nuclear division events such as mitosis or meiosis, where DNA must be compacted and segregated to daughter cells. Condensins and cohesins are *Structural Maintenance of*

Chromosome proteins that aid in the condensation of sister chromatids and the linkage of the centromere in sister chromatids. These SMC proteins induce positive supercoils.

Supercoiling is also required for DNA/RNA synthesis. Because DNA must be unwound for DNA/RNA polymerase action, supercoils will result. The region ahead of the polymerase complex will be unwound; this stress is compensated with positive supercoils ahead of the complex. Behind the complex, DNA is rewound and there will be compensatory negative supercoils. It is important to note that topoisomerases such as DNA gyrase (Type II Topoisomerase) play a role in relieving some of the stress during DNA/RNA synthesis.

Modelling using Mathematics

DNA supercoiling can be described numerically by changes in the 'linking number' Lk. The linking number is the most descriptive property of supercoiled DNA. Lk_0, the number of turns in the relaxed (B type) DNA plasmid/molecule, is determined by dividing the total base pairs of the molecule by the relaxed bp/turn which, depending on reference is 10.4-10.5.

$$Lk_o = bp \,/\, 10.4$$

Lk is merely the number of crosses a single strand makes across the other in a planar projection. The topology of the DNA is described by the equation below in which the linking number is equivalent to the sum of TW, which is the number of twists or turns of the double helix, and Wr which is the number of coils or 'writhes'. If there is a closed DNA molecule, the sum of TW and Wr, or the linking number, does not change. However, there may be complementary changes in TW and Wr without changing their sum.

$$Lk = Tw + Wr$$

The change in the linking number, ΔLk, is the actual number of turns in the plasmid/molecule, Lk, minus the number of turns in the relaxed plasmid/molecule Lk_0.

$$\Delta Lk = Lk - Lk_o$$

If the DNA is negatively supercoiled $\Delta Lk < 0$. The negative supercoiling implies that the DNA is underwound. A standard expression independent of the molecule size is the "specific linking difference" or "superhelical density" denoted σ. σ represents the number of turns added or removed relative to the total number of turns in the relaxed molecule/plasmid, indicating the level of supercoiling.

$$\sigma = \Delta Lk \,/\, Lk_o$$

The Gibbs free energy associated with the coiling is given by the equation below

$\Delta G / N = 10RT\sigma^2$

Examples

Since the linking number L of supercoiled DNA is the number of times the two strands are intertwined (and both strands remain covalently intact), L cannot change. The reference state (or parameter) L_0 of a circular DNA duplex is its relaxed state. In this state, its writhe $W = 0$. Since $L = T + W$, in a relaxed state $T = L$. Thus, if we have a 400 bp relaxed circular DNA duplex, $L \sim 40$ (assuming ~10 bp per turn in B-DNA). Then $T \sim 40$.

Positively supercoiling:

T = 0, W = 0, then L = 0

T = +3, W = 0, then L = +3

T = +2, W = +1, then L = +3

Negatively supercoiling:

T = 0, W = 0, then L = 0

T = -3, W = 0, then L = -3

T = -2, W = -1, then L = -3

Negative supercoils favour local unwinding of the DNA, allowing processes such as transcription, DNA replication, and recombination. Negative supercoiling is also thought to favour the transition between B-DNA and Z-DNA, and moderate the interactions of DNA binding proteins involved in gene regulation.

Molecular Structure of Nucleic Acids: A Structure for Deoxyribose Nucleic Acid

The *Molecular Structure of Nucleic Acids: A Structure for Deoxyribose Nucleic Acid* was an article published by James D. Watson and Francis Crick in the scientific journal *Nature* in its 171st volume on page 737-738 (dated April 25, 1953). It was the first publication which described the discovery of the double helix structure of DNA. This discovery had a major impact on biology, particularly in the field of genetics.

This article is often termed a "pearl" of science because it is brief and contains the answer to a fundamental mystery about living organisms. This mystery was the question of how it was possible that genetic instructions were held inside organisms and how they were passed from generation to generation. The article presents a simple and elegant solution, which surprised many biologists at the time who felt that DNA transmission was going to be more difficult to detail and understand.

The Nature of the Discovery

Watson and Crick's 1953 article contains the answer to a fundamental mystery about living organisms. The nature of their discovery was distinctive and in some ways unsurprising.

What is hidden in the technical jargon of the title is that Watson and Crick's discovery of the chemical structure of DNA finally revealed how genetic instructions are stored inside organisms and passed from generation to generation. However, Watson and Crick took the majority of their information from a little known scientist named Rosalind Franklin, without her knowledge. She discovered the exact structure of DNA before either of the two, however she died before the public could find out about her. Watson never and to this day, has not given her credit for the discovery.

Origins of Molecular Biology

The application of physics and chemistry to biological problems led to the development of molecular biology. Not all biology that concerns molecules falls into the category that is labelled "molecular biology".

Molecular biology is particularly concerned with the flow and consequences of biological information at the level of genes and proteins. The discovery of the DNA double helix made clear that genes are functionally defined parts of DNA molecules and that there must be a way for cells to make use of their DNA genes in order to make proteins.

Linus Pauling was a chemist who was very influential in developing an understanding of the structure of biological molecules. In 1951, Pauling published the structure of the alpha helix, a fundamentally important structural component of proteins. Early in 1953 Pauling published an incorrect triple helix model of DNA. Both Crick, and particularly Watson, felt that they were racing against Pauling to discover the structure of DNA.

Max Delbruck was a physicist who recognized some of the biological implications of quantum physics. Delbruck's thinking about the physical basis of life stimulated Erwin Schrodinger to write the highly influential book, *What Is Life?* Schrodinger's book was an important influence on Francis Crick, James D. Watson, and Maurice Wilkins who won a Nobel prize in Medicine for the discovery of the DNA double helix. Delbruck's efforts to promote the "Phage Group" (exploring genetics by way of the viruses that infect bacteria) was important in the early development of molecular biology in general and the development of Watson's scientific interests in particular.

DNA Structure and Function

It is not always the case that the structure of a molecule is easy to relate to its function. What makes the structure of DNA so obviously related to its function was described modestly at the end of the article: "It has not escaped our notice that the specific pairing we have postulated immediately suggests a possible copying mechanism for the genetic material".

The "specific pairing" is a key feature of the Watson and Crick model of DNA, the pairing of nucleotide subunits. In DNA, the amount of guanine is equal to cytosine and the amount of adenine is equal to thymine. The A:T and C:G pairs are structurally similar. In particular, the length of each base pair is the same and they fit equally between the two phosphate backbones. The base pairs are held together by hydrogen bonds, a type of chemical attraction that is easy to break and easy to reform. After realizing the structural similarity of the A:T and C:G pairs, Watson and Crick soon produced their double helix model of DNA with the hydrogen bonds at the core of the helix providing a way to unzip the two complementary strands for easy replication: the last key requirement for a likely model of the genetic molecule.

Indeed, the base-pairing did suggest a way to copy a DNA molecule. Just pull apart the two phosphate backbones, each with its hydrogen bonded A, T, G, and C components. Each strand could then be used as a template for assembly of a new base-pair complementary strand.

Future Considerations

When Watson and Crick produced their double helix model of DNA, it was known that most of the specialized features of the many different life forms on Earth are made possible by proteins. Structurally, proteins are long chains of amino acid subunits. In some way, the genetic molecule, DNA, had to contain instructions for how to make the thousands of proteins found in cells. From the DNA double helix model, it was clear that there must be some correspondence between the linear sequences of nucleotides in DNA molecules to the linear sequences of amino acids in proteins. The details of how sequences of DNA instruct cells to make specific proteins was worked out by molecular biologists during the period from 1953 to 1965. Francis Crick played an integral role in both the theory and analysis of the experiments that led to an improved understanding of the genetic code.

Consequences

Other advances in molecular biology stemming from the discovery of the DNA double helix eventually led to ways to sequence genes. James Watson played an important role in getting government funding for the Human Genome Project. The ability to sequence and manipulate DNA is now central to the biotechnology industry and modern medicine. The austere beauty of the structure and the practical implications of the DNA double helix combined to make *Molecular structure of Nucleic Acids; A Structure for Deoxyribose Nucleic Acid* one of the most prominent biology articles of the twentieth century.

Controversy

Watson and Crick based their molecular model of the DNA double helix on data that had been collected by researchers in several other laboratories. Watson and Crick were the first to put together all of the scattered fragments of information that were required to produce a successful molecular model of DNA.

Much of the data that were used by Crick and Watson came from unpublished work by Maurice Wilkins, Rosalind Franklin, A.R. Stokes, and H.R. Wilson at King's College London in the University of London. Key data from Wilkins, Stokes, and Wilson, and, separately, by Franklin and Gosling, were published in two separate additional articles in the same issue of *Nature* with the article by Watson and Crick. The article by Watson and Crick did acknowledge that they had been "stimulated" by experimental results from the King's College researchers, and a similar acknowledgment was published by M. H. F. Wilkins, A.R. Stokes, and H. R. Wilson in the following three-page article.

In 1968, Watson published a highly controversial autobiographical account of the discovery of the double-helical, molecular structure of DNA called *The Double Helix*, and which was not accepted—at least publicly—either by Francis Crick or by M.H.F. Wilkins. Furthermore, Erwin Chargaff also printed a rather "unsympathetic review" of James D. Watson's booklet in the March 29, 1968 issue of *Science*. In his `autobiographical' booklet, Watson stated among other things that he and Crick had access to some of Franklin's data from a source that she was not aware of, and also that he had seen—without her permission—the B-DNA X-ray diffraction pattern obtained by Franklin and Gosling in May 1952 at King's in London. In particular, in late 1952, Dr. Franklin had submitted a progress report to the Medical Research Council, which was reviewed by Dr. Max Perutz, then at The Cavendish Laboratory of the University of Cambridge, UK. Watson and Crick also worked in the MRC-supported Cavendish Laboratory in Cambridge whereas Drs. Wilkins and Franklin were in the MRC supported laboratory at King's in London. Such MRC reports were not usually widely circulated, but Crick read a copy of Dr. Franklin's research summary in early 1953.

Max Perutz's justification for passing this information to both Crick and Watson was that the report contained information which Watson has previously heard in November 1951 when Dr. Franklin talked about her unpublished results with Raymond Gosling during a meeting arranged by Dr. M.H.F. Wilkins at King's College, following a request from Crick and Watson; this justification does not hold however for Crick who was not present at this November 1951 meeting, but who also was given access by Max Perutz to Franklin's MRC report data which prompted Crick and Watson to seek permission from Sir Lawrence Bragg—who was at the time the head of the Cavendish Laboratory in Cambridge—to publish in *Nature* their double-helix molecular model of DNA based on Franklin's and also Wilkins's data. Moreover, in November 1951 Watson had acquired—by his own admission—little training in X-ray crystallography, and therefore had not fully understood (again, according to his own admission, in "The Double Helix") what Dr. Franklin was saying about the structural symmetry of the DNA molecule.

Crick, however, knowing the Fourier transforms of Bessel functions that represent the X-ray diffraction patterns of helical structures of atoms, correctly interpreted further one of Dr.Franklin's experimental findings as indicating that DNA was most likely to be a double helix with the two polynucleotide chains running in opposite directions. Crick was thus in a unique position to make this interpretation because he had previously worked on the X-ray diffraction data for other large molecules that had similar, helical symmetry to that of DNA. Dr. Franklin, on the other hand, rejected at first the molecular model building approach proposed by Crick and Watson because their first DNA model presented by Watson to her and Dr. M.H.F. Wilkins in 1952 in London had an obviously incorrect structure with hydrated charged groups on the inside of the model, rather than on the outside, as explicitly admitted by James D. Watson in his "Double Helix" booklet. It is therefore questioned whether Crick's colleague, Dr.Max Perutz, acted unethically by allowing Crick access to Dr. Franklin's MRC report about the crystallographic unit of the B-DNA and A-DNA structures. Dr. Perutz claimed, however, that he felt he had not because this report was not confidential, and had been designed as part of an effort to promote contact between different MRC research groups.

Molecular Models of DNA

Molecular models of DNA structures are representations of the molecular geometry and topology of Deoxyribonucleic acid (DNA) molecules using one of several means, with the aim of simplifying and presenting the essential, physical and chemical, properties of DNA molecular structures either *in vivo* or *in vitro*. These representations include closely packed spheres (CPK models) made of plastic, metal wires for 'skeletal models', graphic computations and animations by computers, artistic rendering. Computer molecular models also allow animations and molecular dynamics simulations that are very important for understanding how DNA functions *in vivo*.

The more advanced, computer-based molecular models of DNA involve molecular dynamics simulations as well as quantum mechanical computations of vibro-rotations, delocalized molecular orbitals (MOs), electric dipole moments, hydrogen-bonding, and so on. DNA molecular dynamics modelling involves simulations of DNA molecular geometry and topology changes with time as a result of both intra- and inter- molecular interactions of DNA. Whereas molecular models of Deoxyribonucleic acid (DNA) molecules such as closely packed spheres (CPK models) made of plastic or metal wires for 'skeletal models' are useful representations of static DNA structures, their usefulness is very limited for representing complex DNA dynamics. Computer molecular modelling allows both animations and molecular dynamics simulations that are very important for understanding how DNA functions *in vivo*.

History

History of Molecular Biology: The history of molecular biology begins in the 1930s with the convergence of various, previously distinct biological disciplines: biochemistry, genetics, microbiology, and virology. With the hope of understanding life at its most fundamental level, numerous physicists and chemists also took an interest in what would become molecular biology.

In its modern sense, molecular biology attempts to explain the phenomena of life starting from the macromolecular properties that generate them. Two categories of macromolecules in particular are the focus of the molecular biologist: 1) nucleic acids, among which the most famous is deoxyribonucleic acid (or DNA), the constituent of genes, and 2) proteins, which are the active agents of living organisms. One definition of the scope of molecular biology therefore is to characterize the structure, function and relationships between these two types of macromolecules. This relatively limited definition will suffice to allow us to establish a date for the so-called "molecular revolution", or at least to establish a chronology of its most fundamental developments.

General Overview

In its earliest manifestations, molecular biology—the name was coined by Warren Weaver of the Rockefeller Foundation in 1938—was an ideal of physical and chemical explanations of life, rather than a coherent discipline. Following the advent of the Mendelian-chromosome theory of heredity in the 1910s and the maturation of atomic theory and quantum mechanics in the 1920s, such explanations seemed within reach. Weaver and others encouraged (and funded) research at the intersection of biology, chemistry and physics, while prominent physicists such as Niels Bohr and Erwin Schrodinger turned their attention to biological speculation. However, in the 1930s and 1940s it was by no means clear which—if any—cross-disciplinary research would bear fruit; work in colloid chemistry, biophysics and radiation biology, crystallography, and other emerging fields all seemed promising.

In 1940, George Beadle and Edward Tatum demonstrated the existence of a precise relationship between genes and proteins. In the course of their experiments connecting genetics with biochemistry, they switched from the genetics mainstay *Drosophila* to a more appropriate model organism, the fungus *Neurospora*; the construction and exploitation of new model organisms would become a recurring theme in the development of molecular biology. In 1944, Oswald Avery, working at the Rockefeller Institute of New York, demonstrated that genes are made up of DNA. In 1952, Alfred Hershey and Martha Chase confirmed that the genetic material of the bacteriophage, the virus which infects bacteria, is made up of DNA. In 1953, James Watson and Francis Crick discovered the double helical structure of the DNA molecule. In 1961, Francois Jacob and Jacques Monod hypothesized the existence of an intermediary between DNA and its protein products, which they called messenger RNA.

Between 1961 and 1965, the relationship between the information contained in DNA and the structure of proteins was determined: there is a code, the genetic code, which creates a correspondence between the succession of nucleotides in the DNA sequence and a series of amino acids in proteins. At the beginning of the 1960s, Monod and Jacob also demonstrated how certain specific proteins, called regulative proteins, latch onto DNA at the edges of the genes and control the transcription of these genes into messenger RNA; they direct the "expression" of the genes.

The chief discoveries of molecular biology took place in a period of only about twenty-five years. Another fifteen years were required before new and more sophisticated technologies, united today under the name of genetic engineering, would permit the isolation and characterization of genes, in particular those of highly complex organisms.

The Exploration of the Molecular Dominion

If we evaluate the molecular revolution within the context of biological history, it is easy to note that it is the culmination of a long process which began with the first observations through a microscope. The aim of these early researchers was to understand the functioning of living organisms by describing their organization at the microscopic level.

From the end of the 18th century, the characterization of the chemical molecules which make up living beings gained increasingly greater attention, along with the birth of physiological chemistry in the 19th century, developed by the German chemist Justus von Liebig and following the birth of biochemistry at the beginning of the 20th, thanks to another German chemist Eduard Buchner. Between the molecules studied by chemists and the tiny structures visible under the optical microscope, such as the cellular nucleus or the chromosomes, there was an obscure zone, "the world of the ignored dimensions," as it was called by the chemical-physicist Wolfgang Ostwald. This world is populated by colloids, chemical compounds whose structure and properties were not well defined.

The successes of molecular biology derived from the exploration of that unknown world by means of the new technologies developed by chemists and physicists: X-ray diffraction, electron microscopy, ultracentrifugization, and electrophoresis. These studies revealed the structure and function of the macromolecules.

A milestone in that process was the work of Dr. Linus Pauling in 1949, which for the first time linked the specific genetic mutation in patients with sickle cell disease to a demonstrated change in an individual protein, the hemoglobin in the erythrocytes of heterozygous or homozygous individuals.

The Encounter between Biochemistry and Genetics

The development of molecular biology is also the encounter of two disciplines which made considerable progress in the course of the first thirty years of the twentieth century: biochemistry and genetics. The first studies the structure and function of the molecules which make up living things. Between 1900 and 1940, the central processes of metabolism were described: the process of digestion and the absorption of the nutritive elements derived from alimentation, such as the sugars. Every one of these processes is catalysed by a particular enzyme. Enzymes are proteins, like the antibodies present in blood or the proteins responsible for muscular contraction. As a consequence, the study of proteins, of their structure and synthesis, became one of the principal objectives of biochemists.

The second discipline of biology which developed at the beginning of the 20th century is genetics. After the rediscovery of the laws of Mendel through the studies of Hugo de Vries, Carl Correns and Erich von Tschermak in 1900, this science began to take shape thanks to the adoption by Thomas Hunt Morgan, in 1910, of a model organism for genetic studies, the famous fruit fly (Drosophila melanogaster). Shortly after, Morgan showed that the genes are localized on chromosomes. Following this discovery, he continued working with Drosophila and, along with numerous other research groups, confirmed the importance of the gene in the life and development of organisms. Nevertheless, the chemical nature of genes and their mechanisms of action remained a mystery. Molecular biologists committed themselves to the determination of the structure, and the description of the complex relations between, genes and proteins.

The development of molecular biology was not just the fruit of some sort of intrinsic "necessity" in the history of ideas, but was a characteristically historical phenomenon, with all of its unknowns, imponderables and contingencies: the remarkable developments in physics at the beginning of the 20th century highlighted the relative lateness in development in biology, which became the "new frontier" in the search for knowledge about the empirical world. Moreover, the developments of the theory of information and cybernetics in the 1940s, in response to military exigencies, brought to the new biology a significant number of fertile ideas and, especially, metaphors.

The choice of bacteria and of its virus, the bacteriophage, as models for the study of the fundamental mechanisms of life was almost natural - they are the smallest living organisms known to exist - and at the same time the fruit of individual choices. This model owes its success, above all, to the fame and the sense of organization of Max Delbruck, a German physicist, who was able to create a dynamic research group, based in the United States, whose exclusive scope was the study of the bacteriophage: the *School of the Phage*.

The geographic panorama of the developments of the new biology was conditioned above all by preceding work. The US, where genetics had developed the most rapidly, and the UK, where there was a coexistence of both genetics and biochemical research of highly advanced levels, were in the avant-garde. Germany, the cradle of the revolutions in physics, with the best minds and the most advanced laboratories of genetics in the world, should have had a primary role in the development of molecular biology. But history decided differently: the arrival of the Nazis in 1933 - and, to a less extreme degree, the rigidification of totalitarian measures in fascist Italy - caused the emigration of a large number of Jewish and non-Jewish scientists. The majority of them fled to the US or the UK, providing an extra impulse to the scientific dynamism of those nations. These movements ultimately made molecular biology a truly international science from the very beginnings.

History of DNA Biochemistry

The study of DNA is a central part of molecular biology.

First Isolation of DNA

Working in the 19th century, biochemists initially isolated DNA and RNA (mixed together) from cell nuclei. They were relatively quick to appreciate the polymeric nature of their "nucleic acid" isolates, but realized only later that nucleotides were of two types—one containing ribose and the other deoxyribose.

It was this subsequent discovery that led to the identification and naming of DNA as a substance distinct from RNA.

Friedrich Miescher (1844–1895) discovered a substance he called "nuclein" in 1869. Somewhat later, he isolated a pure sample of the material now known as DNA from the sperm of salmon, and in 1889 his pupil, Richard Altmann, named it "nucleic acid". This substance was found to exist only in the chromosomes.

In 1919 Phoebus Levene at the Rockefeller Institute identified the components (the four bases, the sugar and the phosphate chain) and he showed that the components of DNA were linked in the order phosphate-sugar-base. He called each of these units a nucleotide and suggested the DNA molecule consisted of a string of nucleotide units linked together through the phosphate groups, which are the 'backbone' of the molecule. However Levene thought the chain was short and that the bases repeated in the same fixed order. Torbjorn Caspersson and Einar Hammersten showed that DNA was a polymer.

Chromosomes and Inherited Traits

Max Delbruck, Nikolai V. Timofeeff-Ressovsky, and Karl G. Zimmer published results in 1935 suggesting that chromosomes are very large molecules the structure of which can be changed by treatment with X-rays, and that by so changing their structure it was possible to change the heritable characteristics governed by those chromosomes. In 1937 William Astbury produced the first X-ray diffraction patterns from DNA. He was not able to propose the correct structure but the patterns showed that DNA had a regular structure and therefore it might be possible to deduce what this structure was.

In 1943, Oswald Theodore Avery and a team of scientists discovered that traits proper to the "smooth" form of the *Pneumococcus* could be transferred to the "rough" form of the same bacteria merely by making the killed "smooth" (S) form available to the live "rough" (R) form. Quite unexpectedly, the living R *Pneumococcus* bacteria were transformed into a new strain of the S form, and the transferred S characteristics turned out to be heritable. Avery called the medium of transfer of traits the transforming principle; he identified DNA as the transforming principle, and not protein as previously thought. He essentially redid Frederick Griffith's experiment. In 1953, Alfred Hershey and Martha Chase did an experiment (Hershey-Chase experiment) that showed, in T2 phage, that DNA is the genetic material (Hershey shared the Nobel prize with Luria).

Discovery of the Structure of DNA

In the 1950s, three groups made it their goal to determine the structure of DNA. The first group to start was at King's College London and was led by Maurice Wilkins and was later joined by Rosalind Franklin. Another group consisting of Francis Crick and James D. Watson was at Cambridge. A third group was at Caltech and was led by Linus Pauling. Crick and Watson built physical models using metal rods and balls, in which they incorporated the known chemical structures of the nucleotides, as well as the known position of the linkages joining one nucleotide to the next along the polymer. At King's College Maurice Wilkins and Rosalind Franklin examined X-ray diffraction patterns of DNA fibres. Of the three groups, only the London group was able to produce good quality diffraction patterns and thus produce sufficient quantitative data about the structure.

Helix Structure

In 1948 Pauling discovered that many proteins included helical shapes. Pauling had deduced this structure from X-ray patterns and from attempts to physically model the structures. (Pauling was also later to suggest an incorrect three chain helical DNA structure based on Astbury's data.) Even in the initial diffraction data from DNA by Maurice Wilkins, it was evident that the structure involved helices. But this insight was only a beginning. There remained the questions of how many strands came together, whether this number was the same for every helix, whether the bases pointed toward the helical axis or away, and ultimately what were the explicit angles and coordinates of all the bonds and atoms. Such questions motivated the modelling efforts of Watson and Crick.

Complementary Nucleotides

In their modelling, Watson and Crick restricted themselves to what they saw as chemically and biologically reasonable. Still, the breadth of possibilities was very wide. A breakthrough occurred in 1952, when Erwin Chargaff visited Cambridge and inspired Crick with a description of experiments Chargaff had published in 1947. Chargaff had observed that the proportions of the four nucleotides vary between one DNA sample and the next, but that for particular pairs of nucleotides — adenine and thymine, guanine and cytosine — the two nucleotides are always present in equal proportions. Using X-ray diffraction, as well as other data from Rosalind Franklin and her information that the bases were paired, James D. Watson and Francis Crick arrived at the first accurate model of DNA's molecular structure in 1953, which was accepted through inspection by Rosalind Franklin.

The discovery was made on February 28, 1953; the first Watson/Crick paper appeared in Nature on April 25, 1953. Sir Lawrence Bragg, the director of the Cavendish Laboratory, where Watson and Crick worked, gave a talk at Guys Hospital Medical School in London on Thursday, May 14, 1953 which resulted in an article by Ritchie Calder in The News Chronicle of London, on

Friday, May 15, 1953, entitled "Why You Are You. Nearer Secret of Life." The news reached readers of The New York Times the next day; Victor K. McElheny, in researching his biography, "Watson and DNA: Making a Scientific Revolution", found a clipping of a six-paragraph New York Times article written from London and dated May 16, 1953 with the headline "Form of `Life Unit' in Cell Is Scanned." The article ran in an early edition and was then pulled to make space for news deemed more important. (The New York Times subsequently ran a longer article on June 12, 1953). The Cambridge University undergraduate newspaper also ran its own short article on the discovery on Saturday, May 30, 1953. Bragg's original announcement at a Solvay conference on proteins in Belgium on 8 April 1953 went unreported by the press! In 1962 Watson, Crick, and Maurice Wilkins jointly received the Nobel Prize for Physiology or Medicine for their determination of the structure of DNA.

"Central Dogma"

Watson and Crick's model attracted great interest immediately upon its presentation. Arriving at their conclusion on February 21, 1953, Watson and Crick made their first announcement on February 28. In an influential presentation in 1957, Crick laid out the "Central Dogma", which foretold the relationship between DNA, RNA, and proteins, and articulated the "sequence hypothesis." A critical confirmation of the replication mechanism that was implied by the double-helical structure followed in 1958 in the form of the Meselson-Stahl experiment. Work by Crick and coworkers showed that the genetic code was based on non-overlapping triplets of bases, called codons, and Har Gobind Khorana and others deciphered the genetic code not long afterward (1966). These findings represent the birth of molecular biology.

History of RNA Tertiary Structure

Pre-history: The Helical Structure of RNA

The earliest work in RNA structural biology coincided, more or less, with the work being done on DNA in the early 1950s. In their seminal 1953 paper, Watson and Crick suggested that van der Waals crowding by the 2'OH group of ribose would preclude RNA from adopting a double helical structure identical to the model they proposed - what we now know as B-form DNA. This provoked questions about the three dimensional structure of RNA: could this molecule form some type of helical structure, and if so, how?

As with DNA, early structural work on RNA centered around isolation of native RNA polymers for fibre diffraction analysis. In part because of heterogeneity of the samples tested, early fibre diffraction patterns were usually ambiguous and not readily interpretable. In 1955, Grunberg-Manago *et al.* published a paper describing the enzyme polynucleotide phosphorylase, which cleaved a phosphate group from nucleotide diphosphates to catalyse their polymerization. This discovery allowed researchers to synthesize homogenous nucleotide polymers, which they then combined to produce double stranded molecules.

These samples yielded the most readily interpretable fibre diffraction patterns yet obtained, suggesting an ordered, helical structure for cognate, double stranded RNA that differed from that observed in DNA. These results paved the way for a series of investigations into the various properties and propensities of RNA. Through the late 1950s and early 1960s, numerous papers were publihed on various topics in RNA structure, including RNA-DNA hybridization, triple stranded RNA, and even small-scale crystallography of RNA di-nucleotides - G-C, and A-U - in primitive helix-like arrangements. For a more in-depth review of the early work in RNA structural biology.

The Beginning: Crystal Structure of tRNA

In the mid-1960s, the role of tRNA in protein synthesis was being intensively studied. At this point, ribosomes had been implicated in protein synthesis, and it had been shown that an mRNA strand was necessary for the formation of these structures. In a 1964 publication, Warner and Rich showed that ribosomes active in protein synthesis contained tRNA molecules bound at the A and P sites, and discussed the notion that these molecules aided in the peptidyl transferase reaction. However, despite considerable biochemical characterization, the structural basis of tRNA function remained a mystery.

In 1965, Holley *et al.* purified and sequenced the first tRNA molecule, initially proposing that it adopted a cloverleaf structure, based largely on the ability of certain regions of the molecule to form stem loop structures. The isolation of tRNA proved to be the first major windfall in RNA structural biology. Following Holley's publication, numerous investigators began work on isolation tRNA for crystallographic study, developing improved methods for isolating the molecule as they worked.

By 1968 several groups had produced tRNA crystals, but these proved to be of limited quality and did not yield data at the resolutions necessary to determine structure. In 1971, Kim *et al.* achieved another breakthrough, producing crystals of yeast tRNA that diffracted to 2-3 Angstrom resolutions by using spermine, a naturally occurring polyamine, which bound to and stabilized the tRNA.

Despite having suitable crystals, however, the structure of tRNA was not immediately solved at high resolution; rather it took pioneering work in the use of heavy metal derivatives and a good deal more time to produce a high-quality density map of the entire molecule. In 1973, Kim *et al.* produced a 4 Angstrom map of the tRNA molecule in which they could unambiguously trace the entire backbone. This solution would be followed by many more, as various investigators worked to refine the structure and thereby more thoroughly elucidate the details of base pairing and stacking interactions, and validate the published architecture of the molecule.

The tRNA structure is notable in the field of nucleic acid structure in general, as it represented the first solution of a long-chain nucleic acid structure of any kind - RNA or DNA - preceding Dickerson's solution of a B-form dodecamer by nearly a decade. Also, tRNA demonstrated many of the tertiary interactions observed in RNA architecture which would not be categorized and more thoroughly understood for years to come, providing a foundation for all future RNA structural research.

The Renaissance: the Hammerhead Ribozyme and the Group I Intron: $P_{4\text{-}6}$

For a considerable time following the first tRNA structures, the field of RNA structure did not dramatically advance. The ability to study an RNA strcuture depended upon the potential to isolate the RNA target. This proved limiting to the field for many years, in part owing to the fact that other know targets - i.e. the ribosome - were significantly more difficult to isolate and crystallize.

Further, because other interesting RNA targets had simply not been identified, or were not sufficiently understood to be deemed interesting, there was simply a lack of things to study structurally. As such, for some twenty years following the original publication of the tRNA structure, the structures of only a handful of other RNA targets were solved, with almost all of these belonging to the transfer RNA family. This unfortunate lack of scope would eventually be overcome largely because of two major advancements in nucleic acid research: the identification of ribozymes, and the ability to produce them via *in vitro* transcription.

Subsequent to Tom Cech's publication implicating the *Tetrahymena* group I intron as an autocatalytic ribozyme, and Sidney Altman's report of catalysis by ribonuclease P RNA, several other catalytic RNAs were identified in the late 1980s, including the hammerhead ribozyme. In 1994, McKay *et al.* published the structure of a 'hammerhead RNA-DNA ribozyme-inhibitor complex' at 2.6 Angstrom resolution, in which the autocatalytic activity of the ribozyme was disrupted via binding to a DNA substrate. The conformation of the ribozyme published in this paper was eventually shown to be one of several possible states, and although this particular sample was catalytically inactive, subsequent structures have revealed its active-state architecture. This structure was followed by Doudna's publication of the structure of the P4-P6 domains of the *Tetrahymena* group I intron, a fragment of the ribozyme originally made famous by Cech.

The second clause in the title of this publication - *Principles of RNA Packing* - concisely evinces the value of these two structures: for the first time, comparisons could be made between well described tRNA structures and those of globular RNAs outside the transfer family. This allowed the framework of categorization to be built for RNA tertiary structure. It was now possible to propose the conservation of motifs, folds, and various local stabilizing interactions. For an early review of these structures and their implications.

In addition to the advances being made in global structure determination via crystallography, the early 1990s also saw the implementation of NMR as a powerful technique in RNA structural biology. Coincident with the large-scale ribozyme structures being solved crystallographically, a number of structures of small RNAs and RNAs complexed with drugs and peptides were solved using NMR.

In addition, NMR was now being used to investigate and supplement crystal structures, as exemplified by the determination of an isolated tetraloop-receptor motif structure published in 1997. Investigations such as this enabled a more precise characterization of the base pairing and base stacking interactions which stabilized the global folds of large RNA molecules.

The importance of understanding RNA tertiary structural motifs was prophetically well described by Michel and Costa in their publication identifying the tetraloop motif: "..it should not come as a surprise if self-folding RNA molecules were to make intensive use of only a relatively small set of tertiary motifs. Identifying these motifs would greatly aid modelling enterprises, which will remain essential as long as the crystallization of large RNAs remains a difficult task".

The Modern Era: The Age of RNA Structural Biology

The resurgence of RNA structural biology in the mid-1990s has caused a veritable explosion in the field of nucleic acid structural research. Since the publication of the hammerhead and $P_{4\text{-}6}$ structures, numerous major contributions to the field have been made.

Some of the most noteworthy examples include the structures of the Group I and Group II introns, and the Ribosome. It should be noted that the first three structures were produced using *in vitro* transcription, and that NMR has played a role in investigating partial components of all four structures - testaments to the indispensability of both techniques for RNA research.

Most recently, the 2009 Nobel Prize in Chemistry was awarded to Ada Yonath, Venkatraman Ramakrishnan, and Thomas Steitz for their structural work on the ribosome, demonstrating the prominent role RNA structural biology has taken in modern molecular biology.

History of Protein Biochemistry

First Isolation and Classification

Proteins were recognized as a distinct class of biological molecules in the eighteenth century by Antoine Fourcroy and others. Members of this class (called the "albuminoids", *Eiweisskorper*, or *matières albuminoides*) were recognized by their ability to coagulate or flocculate under various treatments such as heat or acid; well-known examples at the start of the nineteenth century included albumen from egg whites, blood serum albumin, fibrin, and wheat gluten. The similarity between the cooking of egg whites and the curdling of milk was recognized even in ancient times; for example, the name *albumen* for the egg-white protein was coined by Pliny the Elder from the Latin *albus ovi* (egg white).

With the advice of Jons Jakob Berzelius, the Dutch chemist Gerhardus Johannes Mulder carried out elemental analyses of common animal and plant proteins. To everyone's surprise, all proteins had nearly the same empirical formula, roughly $C_{400}H_{620}N_{100}O_{120}$ with individual sulfur and phosphorus atoms. Mulder published his findings in two papers (1837,1838) and hypothesized that there was one basic substance (*Grundstoff*) of proteins, and that it was synthesized by plants and absorbed from them by animals in digestion. Berzelius was an early proponent of this theory and proposed the name "protein" for this substance in a letter dated 10 July 1838

The name protein that I propose for the organic oxide of fibrin and albumin, I wanted to derive from [the Greek word], because it appears to be the primitive or principal substance of animal nutrition.

Mulder went on to identify the products of protein degradation such as the amino acid, leucine, for which he found a (nearly correct) molecular weight of 131 Da.

Purifications and Measurements of Mass

The minimum molecular weight suggested by Mulder's analyses was roughly 9 kDa, hundreds of times larger than other molecules being studied. Hence, the chemical structure of proteins (their primary structure) was an active area of research until 1949, when Fred Sanger sequenced insulin. The (correct) theory that proteins were linear polymers of amino acids linked by peptide bonds was proposed independently and simultaneously by Franz Hofmeister and Emil Fischer at the same conference in 1902. However, some scientists were sceptical that such long macromolecules could be stable in solution. Consequently, numerous alternative theories of the protein primary structure were proposed, e.g., the colloidal hypothesis that proteins were assemblies of small molecules, the cyclol hypothesis of Dorothy Wrinch, the diketopiperazine hypothesis of Emil Abderhalden and the pyrrol/piperidine hypothesis of Troensgard (1942). Most of these theories had difficulties in accounting for the fact that the digestion of proteins yielded peptides and amino acids. Proteins were finally shown to be macromolecules of well-defined composition (and not colloidal mixtures) by Theodor Svedberg using analytical ultracentrifugation. The possibility that some proteins are non-covalent associations of such macromolecules was shown by Gilbert Smithson Adair (by measuring the osmotic pressure of hemoglobin) and, later, by Frederic M. Richards in his studies of ribonuclease S. The mass spectrometry of proteins has long been a useful technique for identifying posttranslational modifications and, more recently, for probing protein structure.

Most proteins are difficult to purify in more than milligram quantities, even using the most modern methods. Hence, early studies focused on proteins that could be purified in large quantities, e.g., those of blood, egg white, various toxins, and digestive/metabolic enzymes obtained from slaughterhouses. Many techniques of protein purification were developed during World War II in a project led by Edwin Joseph Cohn to purify blood proteins to help keep soldiers alive. In the late 1950s, the Armour Hot Dog Co. purified 1 kg (= one million milligrams) of pure bovine pancreatic ribonuclease A and made it freely available to scientists around the world. This generous act made RNase A the main protein for basic research for the next few decades, resulting in several Nobel Prizes.

Protein Folding and First Structural Models

The study of protein folding began in 1910 with a famous paper by Henrietta Chick and C. J. Martin, in which they showed that the flocculation of a protein was composed of two distinct processes: the precipitation of a protein from solution was *preceded* by another process called denaturation, in which the protein became much less soluble, lost its enzymatic activity and became more chemically reactive. In the mid-1920s, Tim Anson and Alfred Mirsky proposed that denaturation was a reversible process, a correct hypothesis that was initially lampooned by some scientists as “unboiling the egg”. Anson also suggested that denaturation was a two-state (“all-or-none”) process, in which one fundamental molecular transition resulted in the drastic changes in solubility, enzymatic activity and chemical reactivity; he further noted that the free energy changes upon denaturation were much smaller than those typically involved in chemical reactions.

In 1929, Hsien Wu hypothesized that denaturation was protein unfolding, a purely conformational change that resulted in the exposure of amino acid side chains to the solvent. According to this (correct) hypothesis, exposure of aliphatic and reactive side chains to solvent rendered the protein less soluble and more reactive, whereas the loss of a specific conformation caused the loss of enzymatic activity. Although considered plausible, Wu's hypothesis was not immediately accepted, since so little was known of protein structure and enzymology and other factors could account for the changes in solubility, enzymatic activity and chemical reactivity. In the early 1960s, Chris Anfinsen showed that the folding of ribonuclease A was fully reversible with no external cofactors needed, verifying the "thermodynamic hypothesis" of protein folding that the folded state represents the global minimum of free energy for the protein.

The hypothesis of protein folding was followed by research into the physical interactions that stabilize folded protein structures. The crucial role of hydrophobic interactions was hypothesized by Dorothy Wrinch and Irving Langmuir, as a mechanism that might stabilize her cyclol structures. Although supported by J. D. Bernal and others, this (correct) hypothesis was rejected along with the cyclol hypothesis, which was disproven in the 1930s by Linus Pauling (among others).

Instead, Pauling championed the idea that protein structure was stabilized mainly by hydrogen bonds, an idea advanced initially by William Astbury (1933). Remarkably, Pauling's incorrect theory about H-bonds resulted in his *correct* models for the secondary structure elements of proteins, the alpha helix and the beta sheet. The hydrophobic interaction was restored to its correct prominence by a famous article in 1959 by Walter Kauzmann on denaturation, based partly on work by Kaj Linderstrom-Lang. The ionic nature of proteins was demonstrated by Bjerrum, Weber and Arne Tiselius, but Linderstrom-Lang showed that the charges were generally accessible to solvent and not bound to each other (1949).

The secondary and low-resolution tertiary structure of globular proteins was investigated initially by hydrodynamic methods, such as analytical ultracentrifugation and flow birefringence. Spectroscopic methods to probe protein structure (such as circular dichroism, fluorescence, near-ultraviolet and infrared absorbance) were developed in the 1950s. The first atomic-resolution structures of proteins were solved by X-ray crystallography in the 1960s and by NMR in the 1980s. As of 2006, the Protein Data Bank has nearly 40,000 atomic-resolution structures of proteins. In more recent times, cryo-electron microscopy of large macromolecular assemblies and computational protein structure prediction of small protein domains are two methods approaching atomic resolution.

Importance

An old standing dynamic problem is how DNA "self-replication" takes place in living cells that should involve transient uncoiling of supercoiled DNA fibres. Although DNA consists of relatively rigid, very large elongated biopolymer molecules called "fibres" or chains (that are made of repeating nucleotide units of four basic types, attached to deoxyribose and phosphate groups), its molecular structure *in vivo* undergoes dynamic configuration changes that involve dynamically attached water molecules and ions. Supercoiling, packing with histones in chromosome structures, and other such supramolecular aspects also involve *in vivo* DNA topology which is even more complex than DNA molecular geometry, thus turning molecular modelling of DNA into an especially challenging problem for both molecular biologists and biotechnologists.

Like other large molecules and biopolymers, DNA often exists in multiple stable geometries (that is, it exhibits conformational isomerism) and configurational, quantum states which are close to each other in energy on the potential energy surface of the DNA molecule.

Such varying molecular geometries can also be computed, at least in principle, by employing *ab initio* quantum chemistry methods that can attain high accuracy for small molecules, although claims that acceptable accuracy can be also achieved for polynuclelotides, as well as DNA conformations, were recently made on the basis of VCD spectral data. Such quantum geometries define an important class of *ab initio* molecular models of DNA whose exploration has barely started especially in connection with results obtained by VCD in solutions. More detailed comparisons with such *ab initio* quantum computations are in principle obtainable through 2D-FT NMR spectroscopy and relaxation studies of polynucleotide solutions or specifically labelled DNA, as for example with deuterium labels.

In an interesting twist of roles, the DNA molecule itself was proposed to be utilized for quantum computing. Both DNA nanostructures as well as DNA 'computing' biochips have been built.

DNA Structure

The structure of DNA shows a variety of forms, both double-stranded and single-stranded. The mechanical properties of DNA, which are directly related to its structure, are a significant problem for cells. Every process which binds or reads DNA is able to use or modify the mechanical properties of DNA for purposes of recognition, packaging and modification. The extreme length (a chromosome may contain a 10 cm long DNA strand), relative rigidity and helical structure of DNA has led to the evolution of histones and of enzymes such as topoisomerases and helicases to manage a cell's DNA. The properties of DNA are closely related to its molecular structure and sequence, particularly the weakness of the hydrogen bonds and electronic interactions that hold strands of DNA together compared to the strength of the bonds within each strand.

Experimental techniques which can directly measure the mechanical properties of DNA are relatively new, and high-resolution visualization in solution is often difficult. Nevertheless, scientists have uncovered large amount of data on the mechanical properties of this polymer, and the implications of DNA's mechanical properties on cellular processes is a topic of active current research.

The DNA found in many cells can be macroscopic in length - a few centimetres long for each human chromosome. Consequently, cells must compact or "package" DNA to carry it within them. In eukaryotes this is carried by spool-like proteins known as histones, around which DNA winds. It is the further compaction of this DNA-protein complex which produces the well known mitotic eukaryotic chromosomes.

Recombinant DNA Technology

Gene Targeting (GT)

Gene targeting (GT) is a key technology for the rational, accurate and safe exploitation of plantsthrough genetic manipulation. Moreover, it offers the potential to completely knockout the expression of target genes or to make specific changes to gene function, objectives that cannot be achieved by conventional transgenesis.

The ability to target DNA integration would permit the locus-specific integration of a transgene into a predetermined site of the host genome, avoiding the accidental inactivation of an endogenous gene localized at the insertion site or the unexpected expression profiles of the transgene itself, the so-called position effect. Systematic isolation and sequencing of genomic DNA flanking the insertion sites (known as FSTs or Flanking Sequence Tag) offers the opportunity to rapidly characterize plants altered in a candidate gene sequence.

This approach is notably most useful in fully sequenced genomes such as in Arabidopsis thaliana. With 125 Mbp and 26,422 genes, the Arabidopsis genome shows very limited synteny with the 420–466 Mbp and 60,000 predicted genes of the rice genome. The recombination machinery has been well conserved throughout evolution, as an essential component of cell survival. In nature, homologous recombination is a DNA maintenance pathway that protects chromosomes against damage affecting both DNA strands, such as double strand breaks (DSBs) or interstrand crosslinks. DSB repair (DSBR) has been one of the most investigated homologous repair pathways.

Recently plastid genome transformation technique has gained prominence due to its better integration and less chances of random spread. The genome propagated by higher plant plastids, the plastome, is typically a double stranded

DNA molecule of 130 to 160 kb. Over one hundred copies of this genome can be present in a single plastid. It is ideally represented as a circular monomer containing 2 inverted repeats, even though reality is more complex since linear and circular multimers have been frequently detected The complete sequence of this highly polyploidy genome is available for about 20 different species of angiosperms.

The first successful transformation of tobacco was performed using as marker a mutant plastid DNA fragment covering the 16SrRNA gene derived from a line resistant to spectinomycin and streptomycin. Major improvements in the selection process were soon obtained with the dominant aadA marker gene, inactivating spectinomycin or streptomycin. When fused to GFP, this marker can be used to track the selection process. Genes encoding resistance to kanamycin, nptII and more recently aphA-6 are also possible options, and could be more appropriate for some species.

There are three different fates for the external DNA to get integrated into the native genome. They are homologous recombination, illegitimate recombination or nonhomologous end joining, and single-strand annealing. Single-strand annealing (SSA), a third path of repair, requires the presence of repeated sequences on both sides of a break. After exonuclease degradation of the 5' ends, repair occurs by annealing of the two complementary sequences, a process leading to the loss of the genetic information contained between these repeats. With respect to the species preferential DSB repair pathway, HR but also IR mediates transgene integration. This second aspect explains the inefficiency of GT in higher plants, which use HR as a minor pathway of repair. Thus, despite the fact that transgene integration processes are still unclear in plants transgenic DNA would be preferentially integrated by end joining whether or not sharing homology within the host genome.

GM Crops

The U.S. Food and Drug Administration (FDA) has found no evidence to indicate that either ordinary plant deoxyribonucleic acid (DNA) or the DNA inserted into plants using bioengineering presents food safety problems. Nor are the small amounts of the newly expressed proteins likely to change dramatically the safety profile of the plant. If safety concerns should arise, however, they would most likely fall into one of three broad categories: allergens, toxins, or anti-nutrients. FDA has extensive experience in evaluating the safety of such substances in food.

It is important to note that the kinds of food safety testing typically conducted by developers of a bioengineered food crop to ensure that their foods meet all applicable requirements of the Food, Drug and Cosmetics Act (FD&C

Act) address these potential concerns. In the event that something unexpected does occur, this testing provides a way to detect such changes at the developmental stage and defer marketing until any concern is resolved.

As Aforementioned, some of the Food Safety Concerns that could Arise Include

Allergens: Foods normally contain many thousands of different proteins. While the majority of proteins do not cause allergic reactions, virtually all known human allergens are proteins. Since genetic engineering can introduce a new protein into a food plant, it is possible that this technique could introduce a previously unknown allergen into the food supply or could introduce a known allergen into a "new" food.

Toxins: It is possible that a new protein, as introduced into a crop as a result of the genetic modification, could cause toxicity.

Anti-nutrients: It is possible that the introduction of anti-nutrients, such as molecules like phytic acid, could reduce essential dietary minerals such as phosphorus. The use of genetic engineering techniques could also result in unintended alterations in the amounts of substances normally found in a food, such as a reduction of Vitamin C or an increase in the concentration of a naturally occurring toxicant in the plant food.

Legal and Regulatory Issues

One important component in ensuring food safety is the U.S. regulatory structure. The FDA regulates bioengineered plant food in conjunction with the United States Department of Agriculture (USDA) and the Environmental Protection Agency (EPA). FDA has authority under the FD&C Act to ensure the safety of all domestic and imported foods for man or animals in the United States market. The exceptions to this are meat, poultry and certain egg products, which are regulated by USDA. The safety of animal drug residues in meat and poultry, however, is regulated by FDA. Pesticides, including those bioengineered into a food crop, are regulated primarily by EPA. USDA's Animal and Plant Health Inspection Service (APHIS) oversees the agricultural and environmental safety of planting and field testing bioengineered plants. Bioengineered foods and food ingredients must adhere to the same standards of safety under the FD&C Act that apply to their conventionally bred counterparts. This means that these products must be as safe as the traditional foods in the market. FDA has the power to remove a food from the market, or sanction those marketing the food if the food poses a risk to public health. It is important to note that the FD&C Act places a legal duty on developers to ensure that the foods they market to consumers are safe and comply with all legal requirements.

Area under the commercialization of genetically modified (GM), often called biotech crops continued to grow for the ninth consecutive year at a sustained double-digit growth rate of 20% in 2004 (James, 2004). The estimated global area of approved GM crops for 2005 was 90.0 million hectares with $4.70 billion global market value—based on the sale price of GM seed plus any technology fees that apply. The global value of the GM crop market is projected a more than $5.0 billion for 2005 (James, 2004). Commercialization of genetically modified (GM) crops continued to grow for the ninth consecutive year.

It reflects the substantial improvements in productivity, the environment, economics, health and social benefits realized by farmers, consumers and society. At the same time the growing controversy over GM food products increased interest in food labelling and identity preservation (IP) of GM crops. Hence, an IP system must be designed to provide assurances that the desired traits are present (or absent) in a product from the seed source, through all steps of production and delivery, to the end user. There are numerous regulatory issues related to GM crops. These include the testing and acceptance of new GM crops for commercial introduction, both domestically and internationally. Nearly every country has different approaches and many have their own regulatory framework, with an intent to prevent cross-contamination of the conventional food and feed industries. IP tracking software is also available in market to ease the burdens associated with precise record-keeping requirements. The economics of IP has been calculated by various scientists depending on different applied IP systems. Niche-marketing opportunities will grow, because of the availability of GM crops and finally, IP of agricultural commodities from GM crops can provide greater choice and value desired by both agricultural producers and consumers.

Based on annual percentage growth in area, of the eight leading GM crop countries, India had the highest percentage year-on-year growth in 2004 with an increase of 400% in Bt cotton area over 2003, followed by Uruguay (200%), Australia (100%), Brazil (66%), China (32%), south Africa (25%), Canada (23%), Argentina (17%) and USA at 11%. India increased its area of approved Bt cotton, introduced only two years ago, from approximately 38,038 hectares in 2002/ 03 to 560,000 hectares in 2004/05 seasons with Bt coverage of 11.65% and approximately 300,000 small farmers benefited from Bt cotton.

Regulatory Issues Related to GM Crops

There are numerous regulatory issues related to GM crops. These include the testing and acceptance of new GM crops for commercial introduction and the introduction of food products containing ingredients from GM crops, both domestically and internationally. Nearly every country has different approaches and many have their own regulatory framework. Regulation is a very dynamic issue with changes being reviewed and proposed in many countries on an ongoing basis.

Numerous regulatory actions are consequently being proposed as governments react to consumer concerns and pressures. Several countries have or have proposed to create new agencies to specifically cover GM crops. Approaches range from cautious acceptance to attempts to ban (growing and even imports), at least for the foreseeable future, all crops and products with GM traits. Each is approaching the testing, introduction, and acceptance of GM crops in its own manner and on its own time schedule.

Abiotic and Biotic Resistance

Coat Protein Mediated Resistance

CP is an important structural protein as it not only protects the viral nucleic acid from degradation, but also plays an important role in virus infection. Its functions includes acquisition and transmission of virus by vectors, cell to cell and long distance spread of the virus in host plants, and for some viruses, it regulates one or more steps of virus replication.Coat Protein (CP) mediated resistance has been demonstrated for 17 groups of viruses, and so far this strategy has shown best promise. CP transgenes have been shown to be effective in preventing or reducing infection and diseases caused by homologous and closely related viruses (Gonsalves et al., 1998). Coat protein-mediated protection has been reported for Tobacco mosaic virus, TMV,Tomato mosaic virus, ToMV, (Sa), Cucumber mosaic virus, CMV, Alfalfa mosaic virus, AlMV, Potato virus X, PVX, Potato virus Y, PVY, Potato leaf roll virus, PLRV, Papaya ringspot virus (PRSV) and a number of other viruses. CP-mediated resistance in Cantaloupe, Papaya, Potato, Squash and Tomato has been tested under the field conditions with fair degree of protection against most of the viruses.

Anti-HIV Agents among Desert Plants

Around 40 million people are affected due to the Human Immuno-deficiency Virus globally. During the past decades, a large number of anti-viral screening experiments on medicinal plant extracts have been reported and have led to the selection of several extracts active towards herpes viruses.

A promising result of a naturally occurring antiherpetic agent was given by n- docosanol (a natural 22 carbon saturated fatty alcohol) which is undergoing phase III clinical trials in patients. Clinical testing of the topical formulation, or systemic administration of drug suspensions has demonstrated a good therapeutic index, since high doses of n- docosanol do not elicit appreciable toxicity. The findings show that natural products are still potential sources in the search for new antiherpatic agents (Hattori et al., 1995). Various plant extracts used in Ayurvedic medicine for inhibitory effects on HIV virus have been studied (Hattori, personal communication). A large number of such plants occur in semi-arid and arid climate of Rajasthan. Acquired immunodeficiency syndrome (AIDS) , the great pandemic of the second half of the 20th Century, is still a threatening disease world wide. Many research approaches are currently aimed at developing novel agents to arrest the replication of HIV through various targets.

These may include the inhibition of reverse transcriptase (RT), protease (PR), membrane fusion and integrase. HIV PR enzyme has been demonstrated to play an essential role in viral replication. A range of HIV PR inhibitors have been designed and applied in clinical trials such as Sanqunavir, Ritonavir and Indinavir. However, the development of drug resistance by virus, irrespective of the target, remains as an overwhelming problem in AIDS chemotherapy. Thus there is great need to search for and develop new and different anti-HIV candidates from plants and natural products are of considerable importance. In search for anti-HIV active agents from natural products, many attempts at screening traditional medicines have been made.

Biotic and Abiotic Stress

Environmental abiotic stress conditions, and especially drought and salinity, are currently the major factors which reduce crop yields worldwide leading to the fact that more than 800 million people are chronically undernourished.. The United Nations Environment Program estimates that approximately 20% of agricultural land and 50% of cropland in the world is saltstressed This salinity, in particular, is an increasing problem and nearly half of the area under irrigation, is at risk to be lost due o building up of salinity. Therefore genetic improvement of salt tolerance has become an urgent need for the future of agriculture in arid and semiarid regions.. One way of solving this problem would be breeding tolerant varieties of crop plants that can be grown on saline soils, but these breeding programes are time consuming and remained elusive . Hence, many metabolic changes are known to occur in plants subjected to salt stress, physiological parameters such as ionic relations have been suggested to be used as tolerance indicators since they can be related to salt tolerance mechanisms.

Biofuel Production

India has over 180 million of wasteland out of which 90 million ha is uncultivable. The degraded and denuded lands arise due to soil erosions as well as secondary salinizations. However Calotropis procera is a potential plant for bioenergy and biofuel production in semi-arid regions of the country because it is able to gronn on such lands. The plant has a growth potential of 2 dry tones to 40 dry tonnes per ha depending on the agroclimatic conditions of its growth. The plant has high level of regeneration potential and could be harvested up to 4 times a year.

The plant yields valuable hydrocarbons which could be converted into diesel substitutes. The bio-diesel derived from Calotropis procera is free from NOx gases, S02 and Suspended Particulate Matter (SPM) and has high cetane value. Due to it's enormous potential for growth under adverse climatic conditions Calotropis procera is suggested as potential plant for bio-diesel production under semi-arid and arid conditions. Jatropha curcas also provides non-edible oil which could be converted into methyl ester with a gain of glycerine. The JME is used as mix in the ratio of 05, to 20 percent blend to fossil fuel diesel and in Germany the use of Rape seed methyl ester is increasing.

Soil Salinity

Soil salinity affects plant production in many parts of the world, particularly on irrigated land. NaCl is the predominant salt in most saline environments. Many crop species are sensitive to high concentrations of salt with negative impacts on agricultural production. Maize (Zea mays L.) is considered a moderately salt-sensitive plant.. Salt resistance of plants is a complex phenomenon that involves biochemical and physiological processes as well as morphological and developmental changes.. In addition to general osmotic stress, high concentrations of Na+ are toxic to maize and molecular mechanisms for salt resistance have not been fully identified or characterized (Zoerb et al. 2004). The analysis of the plant's proteome is an important amendment to the analysis of the genome, because gene expression is altered under salinity stress.

The proteome, in contrast to the genome, is not static but rather dependent on a number of responses influenced by internal and external factors. The plant adaptation to environmental stress, such as soil salinity, is expected to have a strong influence on proteins. One approach to study the molecular mechanisms of plant responses to salinity is to use 2D polyacrylamide gel electrophoresis.

Furthermore, the identification of differentially regulated proteins can lead to the identification of proteins and their corresponding genes which are involved in the physiology of salt resistance. The high resolution achieved by 2D gels and computer-assisted analysis of the differentially regulated proteins were used to examine those proteins whose synthesis was modulated by salt treatment and to quantify these changes. As far as we know, our investigations are the first to characterize the differentially expressed proteins from roots and shoots of maize after treatment with low and high salt stress. Plant material was an efficiently Na+-excluding maize inbred line developed in our laboratory. According to Munns the growth response to salt stress consists of two phases, first, a water-deficit that results from the relatively high solute concentrations in the root medium and, second, ion-specific stress resulting from altered K+/Na+ ratios or Na+ and Cl" concentrations that are toxic to plants. The aim of this study was to elucidate biochemical and physiological reactions of glycophytes to salt stress in the first phase of salinity.

While all major crops, as well as most wild species, are glycophytes, i.e. sensitive to relatively low salt concentrations, there are plants naturally adapted to conditions of high salinity in the soil. These plants, known as halophytes, include a large taxonomic variety and occupy diverse habitats, from extreme dry to temporarily waterlogged sites or salt marshes, and can tolerate NaCl concentrations similar, or even higher than that of sea water, ca. 500 mM.

It is - without doubt - necessary to develop sustainable biological production systems which can tolerate higher water salinity because freshwater resources will not come up with increasing demands of agricultural practice in near future. The sustainable use of halophytic plants is a promising approach to valorize strongly salinised zones unsuitable for conventional agriculture and mediocre waters.

The development of cash crop halophytes and the breeding of salt resistant crop varieties will require a clear understanding of the complex mechanisms of salt stress tolerance, which we are still lacking despite intensive research during the last decade.

Recombinant Proteins

It has been estimated that 1 g of recombinant antibody could be produced in leaves of a plant crop for only about US$100 while the current prices for monoclonal antibodies range from US$2000 to US$5000 per gram. Indeed the cost of producing 1 kg of recombinant protein from most major field crops is estimated to be 10 to 50 times lower than the cost of producing the same amount by E. coli fermentation. Whole plants also have an advantage when tissues such as a fruit, tuber, etc. can be used to express the protein of interest (James and Lee 2001), and an area of undisputed advantage occurs when the oral delivery of pharmaceuticals, as well as feed and food enzymes, is possible. However, there are also some evident obstacles that arise when a whole plant is used for large-scale protein production.

Cell-free systems have proved to have high utility at the genomic, transcriptomic and proteomic levels and to form a vital component of many aspects of recombinant gene expression, and of both structural and functional proteomics..

Compared with DNA microarrays, protein bio-chips provide more challenges and have yet to be perfected due to the complexity and inherent difficulties with protein immobilization.

Novel Cell Free Translation System is Unique Discovery

A novel cell-free translation system is described in which template-mRNA molecules were captured onto solid surfaces to simultaneously synthesize and immobilize proteins in a more native-state form. This technology comprises a novel solid-phase approach to cell-free translation and RNA–protein fusion techniques. A newly constructed biotinylated linker-DNA which enables puromycin-assisted RNA–protein fusion is ligated to the 32 ends of the mRNA molecules to attach the mRNA-template on a streptavidin-coated surface and further to enable the subsequent reactions of translation and RNA–protein fusion on surface.

The protein products are therefore directly immobilized onto solid surfaces and furthermore were discovered to adopt a more native state with proper protein folding and superior biological activity compared with conventional liquid-phase approaches. We further validate this approach via the production of immobilized green fluorescent protein (GFP) on microbeads and by the production and assay of aldehyde reductase (ALR) enzyme with 4-fold or more activity. The approach developed in this study may enable to embrace the concept of the transformation of 'RNA chip-to-protein chip' using a solid-phase cell-free translation system and thus to the development of high-throughput microarray platform in the field of functional genomics and in vitro evolution.

Plant Tissue Culture

Another area of biotechnology is micropropagation of plants. The aim of this technique is a fast production of a great number of genetically identical plants from a highly valuable mother plant or e.g. monosexual male and female plants. These plants can be either directly sold on the market for planting, used for breeding purposes, for genetechnology or the technique is used as a method for basic science studies. Using petiole explants from transgenic plants containing the auxin responsive MAS promoter linked to the GUS reporter gene the distribution of auxin within the cultured petiole could be followed during the induction phase of somatic embryogenesis. Interestingly, the cells forming the glandular canal contain high concentrations of auxins as shown by using transgenic plants containing the auxin sensitive MAS-promoter coupled to the GUS-gene, whatever the significance. Rhizogenic centres develop near vascular bundles prior to those embryogenic centres.

Genetic factors play a central role to induce somatic embryos, i.e. to provide the competence of the species for the process. Here, great variation can be found even within a genera such as Daucus. Eight of twelve Daucus species cultured in identical conditions produced somatic embryos (D. halophilus, D. capillifolius, D. commutatus, D. azoricus, D. gadacei, D. maritimus, D. maximus, D. carota), whereas four species (D. montevidensis, D. pussillus, D. muricatus, D. glochidiatus) were not competent to do so.

Under identical culture conditions, only 8 out of 12 species and subspecies of the genus Daucus proved capable of somatic embryogenesis. Random amplified polymorphic DNA analysis indicated a polymorphism between the genomes of individual species that were capable of embryogenesis and those that were not. Two specific bands (1.1 kbp, 0.68 kbp) were detected only inthe genomes of individuals with the capacity for embryogenesis.

These were cloned and sequenced, and the homology of the nucleotide sequences of the various species was detected: this ranged from 74% to 92% for the larger sequence and from 92% to 97% for the smaller one. These DNA sequences would appear to be useful as a marker of the capacity for somatic embryogenesis in the genus Daucus (Imani et al.2001) The sequences obtained in this study have been registered with the European Bioinformatics Institute (EMBL).

The access numbers for the sequences are: AJ278039 DCA78039; AJ278040 DCA78040; AJ278041 CA78041; AJ278042 DCA78042; AJ278176 DCA278176; AJ278177 DCA278177; AJ278178 DCA278178; AJ278179 DCA278179. No open reading frames were detected.

Later additional studies were performed with other Daucus species (D. capillifolius; D. carota ssp. Azoricus and gadecaei) to determine the use-fulness of these RAPD products as markers for identifying the ability of Daucus species to generate somatic embryos. There was a 100% correlation between the embryogenic potential of the species and the occurrence of the 1.1-kbp and 0.68-kbp band (Imani et al.,2001). Micropropagation Technique in Enhancing the Productivity of Crops have been taken up at large scale at TERI. Some of the activities undertaken at MTP include:

- Large-scale production of superior quality planting material of various economically important plant species using tissue culture technology
- Mass multiplication of those species which are difficult to regenerate by conventional methods of propagation or where conventional methods of propagation are inadequate to meet the demand of planting material
- Development of new micropropagation protocols and refining of others so as to make them suitable for large-scale production of plants
- Helping the entrepreneurs/industry through technology transfer, mother cultures and training of staff

- Assisting clients in setting-up their own tissue culture labs
- Creating awareness.

Till date over 15 million plants of forest species, cash crops, medicinal and aromatic plants, and ornamentals (foliage and flowering) have been dispatched to various state forest and horticulture departments, private entrepreneurs, nurseries, farmers etc. for field demonstrations and routine plantations. In addition, MTP is in possession of micropropagation protocols for over 90 economically important plant species. Field demonstration plots of tissue cultured plants have been laid at different locations to evaluate and compare their growth performances with conventional plants. Besides transfer of technologies to industry for commercialization, MTP has been instrumental in capacity building and creating awareness about tissue culture technology through seminars/ workshops/training programmes, exhibitions, etc.

3

Molecular Cloning

Molecular cloning refers to a set of experimental methods in molecular biology that are used to assemble recombinant DNA molecules and to direct their replication within host organisms. The use of the word *cloning* refers to the fact that the method involves the replication of a single DNA molecule starting from a single living cell to generate a large population of cells containing identical DNA molecules. Molecular cloning generally uses DNA sequences from two different organisms: the species that is the source of the DNA to be cloned, and the species that will serve as the living host for replication of the recombinant DNA. Molecular cloning methods are central to many contemporary areas of modern biology and medicine. In a conventional molecular cloning experiment, the DNA to be cloned is obtained from an organism of interest, then treated with enzymes in the test tube to generate smaller DNA fragments. Subsequently, these fragments are then combined with vector DNA to generate recombinant DNA molecules. The recombinant DNA is then introduced into a host organism (typically an easy-to-grow, benign, laboratory strain of *E. coli* bacteria). This will generate a population of organisms in which recombinant DNA molecules are replicated along with the host DNA. Because they contain foreign DNA fragments, these are transgenic or genetically-modified microorganisms (GMO). This process takes advantage of the fact that a single bacterial cell can be induced to take up and replicate a single recombinant DNA molecule. This single cell can then be expanded exponentially to generate a large amount of bacteria, each of which contain copies of the original recombinant molecule. Thus, both the resulting bacterial population, and the recombinant DNA molecule, are commonly referred to as "clones". Strictly speaking, *recombinant DNA* refers to DNA molecules, while *molecular cloning* refers to the experimental methods used to assemble them.

History of Molecular Cloning

Prior to the 1970s, our understanding of genetics and molecular biology was severely hampered by an inability to isolate and study individual genes from complex organisms. This changed dramatically with the advent of molecular cloning methods. Microbiologists, seeking to understand the molecular mechanisms through which bacteria restricted the growth of bacteriophage, isolated restriction endonucleases, enzymes that could cleave DNA molecules only when specific DNA sequences were encountered. They showed that restriction enzymes cleaved chromosome-length DNA molecules at specific locations, and that specific sections of the larger molecule could be purified by size fractionation. Using a second enzyme, DNA ligase, fragments generated by restriction enzymes could be joined in new combinations, termed recombinant DNA. By recombining DNA segments of interest with vector DNA, such as bacteriophage or plasmids, which naturally replicate inside bacteria, large quantities of purified recombinant DNA molecules could be produced in bacterial cultures. The first recombinant DNA molecules were generated and studied in 1972.

Overview

Molecular cloning takes advantage of the fact that the chemical structure of DNA is fundamentally the same in all living organisms. Therefore, if any segment of DNA from any organism is inserted into a DNA segment containing the molecular sequences required for DNA replication, and the resulting recombinant DNA is introduced into the organism from which the replication sequences were obtained, then the foreign DNA will be replicated along with the host cell's DNA in the transgenic organism.

Molecular cloning is similar to polymerase chain reaction (PCR) in that it permits the replication of a specific DNA sequence. The fundamental difference between the two methods is that molecular cloning involves replication of the DNA in a living microorganism, while PCR replicates DNA in an *in vitro* solution, free of living cells.

Steps in Molecular Cloning

In standard molecular cloning experiments, the cloning of any DNA fragment essentially involves seven steps: (1) Choice of host organism and cloning vector, (2) Preparation of vector DNA, (3) Preparation of DNA to be cloned, (4) Creation of recombinant DNA, (5) Introduction of recombinant DNA into host organism, (6) Selection of organisms containing recombinant DNA, (7) Screening for clones with desired DNA inserts and biological properties.

Choice of Host Organism and Cloning Vector

Although a very large number of host organisms and molecular cloning vectors are in use, the great majority of molecular cloning experiments begin with a laboratory strain of the bacterium *E. coli* (Escherichia coli) and a plasmid cloning vector. *E. coli* and plasmid vectors are in common use because they are technically sophisticated, versatile, widely available, and offer rapid growth of recombinant organisms with minimal equipment. If the DNA to be cloned is exceptionally large (hundreds of thousands to millions of base pairs), then a bacterial artificial chromosome or yeast artificial chromosome vector is often chosen.

Specialized applications may call for specialised host-vector systems. For example, if the experimentalists wish to harvest a particular protein from the recombinant organism, then an expression vector is chosen that contains appropriate signals for transcription and translation in the desired host organism. Alternatively, if replication of the DNA in different species is desired (for example transfer of DNA from bacteria to plants), then a multiple host range vector (also termed shuttle vector) may be selected. In practice, however, specialized molecular cloning experiments usually begin with cloning into a bacterial plasmid, followed by subcloning into a specialized vector.

Whatever combination of host and vector are used, the vector almost always contains four DNA segments that are critically important to its function and experimental utility—(1) an origin of DNA replication is necessary for the vector (and recombinant sequences linked to it) to replicate inside the host organism, (2) one or more unique restriction endonuclease recognition sites that serves as sites where foreign DNA may be introduced, (3) a selectable genetic marker gene that can be used to enable the survival of cells that have taken up vector sequences, and (4) an additional gene that can be used for screening which cells contain foreign DNA.

Preparation of Vector DNA

The cloning vector is treated with a restriction endonuclease to cleave the DNA at the site where foreign DNA will be inserted. The restriction enzyme is chosen to generate a configuration at the cleavage site that is compatible with that at the ends of the foreign DNA. Typically, this is done by cleaving the vector DNA and foreign DNA with the same restriction enzyme, for example EcoRI. Most modern vectors contain a variety of convenient cleavage sites that are unique within the vector molecule (so that the vector can only be cleaved at a single site) and is located within a gene (frequently beta-galactosidase) whose inactivation can be used to distinguish recombinant from non-recombinant organisms at a later step in the process. To improve the ratio of recombinant to non-recombinant organisms, the cleaved vector may be treated with an enzyme (alkaline phosphatase) that modifies the vector ends in such a way that it cannot replicate within cells unless it contains foreign DNA.

Preparation of DNA to be Cloned

For cloning of genomic DNA, the DNA to be cloned is extracted from the organism of interest. Virtually any tissue source can be used (even tissues from extinct animals), as long as the DNA is not extensively degraded. The DNA is then purified using simple methods to remove contaminating proteins (extraction with phenol), RNA (ribonuclease) and smaller molecules (precipitation and/or chromatography). Polymerase chain reaction (PCR) methods are often used for amplification of specific DNA or RNA (RT-PCR) sequences prior to molecular cloning.

DNA for cloning experiments may also be obtained from RNA using reverse transcriptase (complementary DNA or cDNA cloning), or in the form of synthetic DNA (artificial gene synthesis). cDNA cloning is usually used to obtain clones representative of the mRNA population of the cells of interest, while synthetic DNA is used to obtain any precise sequence defined by the designer.

The purified DNA is then treated with a restriction enzyme to generate fragments with ends capable of being linked to those of the vector. If necessary, short double-stranded segments of DNA containing desired restriction sites may be added to create end structures that are compatible with the vector.

Creation of Recombinant DNA with DNA Ligase

The creation of recombinant DNA is in many ways the simplest step of the molecular cloning process. DNA prepared from the vector and foreign source are simply mixed together at appropriate concentrations and exposed to an enzyme (DNA ligase) that covalently links the ends together. This joining reaction is often termed ligation. The resulting DNA mixture containing randomly joined ends is then ready for introduction into the host organism. DNA ligase only recognises and acts on the ends of linear DNA molecules, usually resulting a complex mixture of DNA molecules with randomly joined ends. The desired products (vector DNA covalently linked to foreign DNA) will be present, but other sequences (e.g. foreign DNA linked to itself, vector DNA linked to itself and higher-order combinations of vector and foreign DNA) are also usually present. This complex mixture is sorted out in subsequent steps of the cloning process, after the DNA mixture is introduced into cells.

Introduction of Recombinant DNA into Host Organism

The DNA mixture, previously manipulated in vitro, is moved back into a living cell, referred to as the host organism. The methods used to get DNA into cells are varied, and the name applied to this step in the molecular cloning process will often depend upon the experimental method that is chosen (e.g. transformation, transduction, transfection, electroporation).

When microorganisms are able to take up and replicate DNA from their local environment, the process is termed transformation, and cells that are in a physiological state such that they can take up DNA are said to be competent. In mammalian cell culture, the analogous process of introducing DNA into cells is commonly termed transfection. Both transformation and transfection usually require preparation of the cells through a special growth regime and chemical treatment process that will vary with the specific species and cell types that are used.

Electroporation uses high voltage electrical pulses to translocate DNA across the cell membrane (and cell wall, if present). In contrast, transduction involves the packaging of DNA into virus-derived particles, and using these virus-like particles to introduce the encapsulated DNA into the cell through a process resembling viral infection. Although electroporation and transduction are highly specialised methods, they may be the most efficient methods to move DNA into cells.

Selection of Organisms Containing Vector Sequences

Whichever method is used, the introduction of recombinant DNA into the chosen host organism is usually a low efficiency process; that is, only a small fraction of the cells will actually take up DNA. Experimental scientists deal with this issue through a step of artificial genetic selection, in which cells that have not taken up DNA are selectively killed, and only those cells that can actively replicate DNA containing the selectable marker gene encoded by the vector are able to survive.

When bacterial cells are used as host organisms, the selectable marker is usually a gene that confers resistance to an antibiotic that would otherwise kill the cells, typically ampicillin. Cells harbouring the vector will survive when exposed to the antibiotic, while those that have failed to take up vector sequences will die. When mammalian cells (e.g. human or mouse cells) are used, a similar strategy is used, except that the marker gene confers resistance to the antibiotic Geneticin.

Screening for Clones with Desired DNA Inserts and Biological Properties

Modern bacterial cloning vectors (e.g. pUC19 and later derivatives including the pGEM vectors) use the blue-white screening system to distinguish colonies (clones) of transgenic cells from those that contain the parental vector (i.e. vector DNA with no recombinant sequence inserted). In these vectors, foreign DNA is inserted into a sequence that encodes an essential part of beta-galactosidase, an enzyme whose activity results in formation of a blue-coloured colony on the culture medium that is used for this work. Insertion of the foreign DNA into the beta-galactosidase coding sequence disables the function of the enzyme, so that colonies containing recombinant plasmids remain colourless (white). Therefore, experimentalists are easily able to identify and conduct further studies on transgenic bacterial clones, while ignoring those that do not contain recombinant DNA.

The total population of individual clones obtained in a molecular cloning experiment is often termed a DNA library. Libraries may be highly complex (as when cloning complete genomic DNA from an organism) or relatively simple (as when moving a previously-cloned DNA fragment into a different plasmid), but it is almost always necessary to examine a number of different clones to be sure that the desired DNA construct is obtained. This may be accomplished through a very wide range of experimental methods, including the use of nucleic acid hybridisations, antibody probes, polymerase chain reaction, restriction fragment analysis and/or DNA sequencing.

Applications of Molecular Cloning

Molecular cloning provides scientists with an essentially unlimited quantity of any individual DNA segments derived from any genome. This material can be used for a wide range of purposes, including those in both basic and applied biological science. A few of the more important applications are summarized here.

Genome Organisation and Gene Expression

Molecular cloning has led directly to the elucidation of the complete of sexual DNA sequence of the genomes of a very large number of species and to an exploration of genetic diversity within individual species, work that has been done mostly by determining the DNA sequence of large numbers of randomly cloned fragments of the genome, and assembling the overlapping sequences.

At the level of individual genes, molecular clones are used to generate probes that are used for examining how genes are expressed, and how that expression is related to other processes in biology, including the metabolic environment, extracellular signals, development, learning, senescence and cell death. Cloned genes can also provide tools to examine the biological function and importance of individual genes, by allowing investigators to inactivate the genes, or make more subtle mutations using regional mutagenesis or site-directed mutagenesis.

Production of Recombinant Proteins

Obtaining the molecular clone of a gene can lead to the development of organisms that produce the protein product of the cloned genes, termed a recombinant protein. In practice, it is frequently more difficult to develop an organism that produces an active form of the recombinant protein in desirable quantities than it is to clone the gene. This is because the molecular signals for gene expression are complex and variable, and because protein folding, stability and transport can be very challenging.

Many useful proteins are currently available as recombinant products. These include—(1) medically-useful proteins whose administration can correct a defective or poorly-expressed gene (e.g. recombinant factor VIII, a blood-clotting factor deficient in some forms of hemophilia, and recombinant insulin, used to treat some forms of diabetes), (2) proteins that can be administered to assist in a life threatening emergency (e.g. tissue plasminogen activator, used to treat strokes), and (3) recombinant subunit vaccines, in which a purified protein can be used to immunize patients against infectious diseases, without exposing them to the infectious agent itself (e.g. hepatitis B vaccine).

Transgenic Organisms

Once characterized and manipulated to provide signals for appropriate expression, cloned genes may be inserted into organisms, generating transgenic organisms, also termed genetically-modified organisms (GMOs). Although most GMOs are generated for purposes of basic biological research, a number of GMOs have been developed for commercial use, ranging from animals and plants that produce pharmaceuticals or other compounds (pharming), herbicide-resistant crop plants, and fluorescent tropical fish (GloFish) for home entertainment.

Gene Therapy

Gene therapy involves supplying a functional gene to cells lacking that function, with the aim of correcting a genetic disorder or acquired disease. Gene therapy can be broadly divided into two categories. The first is alteration of germ cells, that is, sperm or eggs, which results in a permanent genetic change for the whole organism and subsequent generations. This “germ line gene therapy” is considered by many to be unethical in human beings. The second type of gene therapy, “somatic cell gene therapy”, is analogous to an organ transplant. In this case, one or more specific tissues are targeted by direct treatment or by removal of the tissue, addition of the therapeutic gene or genes in the laboratory, and return of the treated cells to the patient. Clinical trials of somatic cell gene therapy began in the late 1990s, mostly for the treatment of cancers and blood, liver, and lung disorders.

Despite a great deal of publicity and promises, the history of human gene therapy has been characterised by relatively limited success. The effect of introducing a gene into cells often promotes only partial and/or transient relief from the symptoms of the disease being treated. Some gene therapy trial patients have suffered adverse consequences of the treatment itself, including deaths. In some cases, the adverse effects result from disruption of essential genes within the patient's genome by insertional inactivation. In others, viral vectors used for gene therapy have been contaminated with infectious virus. Nevertheless, gene therapy is still held to be a promising future area of medicine, and is an area where there is a significant level of research and development activity.

Cloning of Viral Antigens

A virus is a small infectious agent that can replicate only inside the living cells of organisms. Viruses infect all types of organisms, from animals and plants to bacteria and archaea. Since Dmitri Ivanovsky's 1892 article describing a non-bacterial pathogen infecting tobacco plants, and the discovery of the tobacco mosaic virus by Martinus Beijerinck in 1898, about 5,000 viruses have been described in detail, although there are millions of different types. Viruses are found in almost every ecosystem on Earth and are the most abundant type of biological entity. The study of viruses is known as virology, a sub-speciality of microbiology.

Virus particles (known as *virions*) consist of two or three parts: the genetic material made from either DNA or RNA, long molecules that carry genetic information; a protein coat that protects these genes; and in so al contact and by exposure to infected blood. The range of host cells that a virus can infect is called its "host range". This can be narrow or, as when a virus is capable of infecting many species, broad. Viral infections in animals provoke an immune response that usually eliminates the infecting virus. Immune responses can also be produced by vaccines, which confer an artificially acquired immunity to the specific viral infection. However, some viruses including those causing AIDS and viral hepatitis evade these immune responses and result in chronic infections. Antibiotics have no effect on viruses, but several antiviral drugs have been developed.

Etymology

The word is from the Latin *virus* referring to poison and other noxious substances, first used in English in 1392. *Virulent*, from Latin *virulentus* (poisonous), dates to 1400. A meaning of "agent that causes infectious disease" is first recorded in 1728, before the discovery of viruses by Dmitry Ivanovsky in 1892. The plural is *viruses*. The adjective *viral* dates to 1948. The term *virion* is also used to refer to a single infective viral particle.

History

Louis Pasteur was unable to find a causative agent for rabies and speculated about a pathogen too small to be detected using a microscope. In 1884, the

French microbiologist Charles Chamberland invented a filter (known today as the Chamberland filter or Chamberland-Pasteur filter) with pores smaller than bacteria. Thus, he could pass a solution containing bacteria through the filter and completely remove them from the solution. In 1892, the Russian biologist Dmitry Ivanovsky used this filter to study what is now known as the tobacco mosaic virus. His experiments showed that crushed leaf extracts from infected tobacco plants remain infectious after filtration. Ivanovsky suggested the infection might be caused by a toxin produced by bacteria, but did not pursue the idea. At the time it was thought that all infectious agents could be retained by filters and grown on a nutrient medium – this was part of the germ theory of disease. In 1898, the Dutch microbiologist Martinus Beijerinck repeated the experiments and became convinced that the filtered solution contained a new form of infectious agent. He observed that the agent multiplied only in cells that were dividing, but as his experiments did not show that it was made of particles, he called it a *contagium vivum fluidum* (soluble living germ) and reintroduced the word *virus*. Beijerinck maintained that viruses were liquid in nature, a theory later discredited by Wendell Stanley, who proved they were particulate. In the same year Friedrich Loeffler and Frosch passed the first animal virus – agent of foot-and-mouth disease (aphthovirus) – through a similar filter.

In the early 20th century, the English bacteriologist Frederick Twort discovered a group of viruses that infect bacteria, now called bacteriophages (or commonly *phages*), and the French-Canadian microbiologist Félix d'Herelle described viruses that, when added to bacteria on agar, would produce areas of dead bacteria. He accurately diluted a suspension of these viruses and discovered that the highest dilutions (lowest virus concentrations), rather than killing all the bacteria, formed discrete areas of dead organisms. Counting these areas and multiplying by the dilution factor allowed him to calculate the number of viruses in the original suspension. Phages were heralded as a potential treatment for diseases such as typhoid and cholera, but their promise was forgotten with the development of penicillin. The study of phages provided insights into the switching on and off of genes, and a useful mechanism for introducing foreign genes into bacteria.

By the end of the 19th century, viruses were defined in terms of their infectivity, their ability to be filtered, and their requirement for living hosts. Viruses had been grown only in plants and animals. In 1906, Ross Granville Harrison invented a method for growing tissue in lymph, and, in 1913, E. Steinhardt, C. Israeli, and R. A. Lambert used this method to grow vaccinia virus in fragments of guinea pig corneal tissue. In 1928, H. B. Maitland and M. C. Maitland grew vaccinia virus in suspensions of minced hens' kidneys. Their method was not widely adopted until the 1950s, when poliovirus was grown on a large scale for vaccine production. Another breakthrough came in 1931, when the American pathologist Ernest William Goodpasture grew influenza and several other viruses in fertilized chickens' eggs. In 1949, John F. Enders, Thomas Weller, and Frederick Robbins grew polio virus in cultured human embryo cells, the first virus to be grown without using solid animal tissue or eggs. This work enabled Jonas Salk to make an effective polio vaccine.

The first images of viruses were obtained upon the invention of electron microscopy in 1931 by the German engineers Ernst Ruska and Max Knoll. In 1935, American biochemist and virologist Wendell Meredith Stanley examined the tobacco mosaic virus and found it was mostly made of protein. A short time later, this virus was separated into protein and RNA parts.

The tobacco mosaic virus was the first to be crystallised and its structure could therefore be elucidated in detail. The first X-ray diffraction pictures of the crystallized virus were obtained by Bernal and Fankuchen in 1941. On the basis of her pictures, Rosalind Franklin discovered the full DNA structure of the virus in 1955. In the same year, Heinz Fraenkel-Conrat and Robley Williams showed that purified tobacco mosaic virus RNA and its coat protein can assemble by themselves to form functional viruses, suggesting that this simple mechanism was probably the means through which viruses were created within their host cells.

The second half of the 20th century was the golden age of virus discovery and most of the 2,000 recognized species of animal, plant, and bacterial viruses were discovered during these years. In 1957, equine arterivirus and the cause of Bovine virus diarrhea (a pestivirus) were discovered. In 1963, the hepatitis B virus was discovered by Baruch Blumberg, and in 1965, Howard Temin described the first retrovirus. Reverse transcriptase, the key enzyme that retroviruses use to translate their RNA into DNA, was first described in 1970, independently by Howard Martin Temin and David Baltimore. In 1983 Luc Montagnier's team at the Pasteur Institute in France, first isolated the retrovirus now called HIV.

Origins

Viruses are found wherever there is life and have probably existed since living cells first evolved. The origin of viruses is unclear because they do not form fossils, so molecular techniques have been used to compare the DNA or RNA of viruses and are a useful means of investigating how they arose. There are three main hypotheses that try to explain the origins of viruses:

Regressive hypothesis : Viruses may have once been small cells that parasitised larger cells. Over time, genes not required by their parasitism were lost. The bacteria rickettsia and chlamydia are living cells that, like viruses, can reproduce only inside host cells. They lend support to this hypothesis, as their dependence on parasitism is likely to have caused the loss of genes that enabled them to survive outside a cell. This is also called the *degeneracy hypothesis*, or *reduction hypothesis*.

Cellular origin hypothesis: Some viruses may have evolved from bits of DNA or RNA that "escaped" from the genes of a larger organism. The escaped DNA could have come from plasmids (pieces of naked DNA that can move *between* cells) or transposons (molecules of DNA that replicate and move around to different positions *within* the genes of the cell). Once called "jumping genes", transposons are examples of mobile genetic elements and could be the origin of some viruses. They were discovered in maize by Barbara McClintock in 1950. This is sometimes called the *vagrancy hypothesis*, or the *escape hypothesis*.

Coevolution hypothesis: This is also called the *virus-first hypothesis* and proposes that viruses may have evolved from complex molecules of protein and nucleic acid at the same time as cells first appeared on earth and would have been dependent on cellular life for billions of years. Viroids are molecules of RNA that are not classified as viruses because they lack a protein coat. However, they have characteristics that are common to several viruses and are often called subviral agents. Viroids are important pathogens of plants. They do not code for proteins but interact with the host cell and use the host machinery for their replication. The hepatitis delta virus of humans has an RNA genome similar to viroids but has a protein coat derived from hepatitis B virus and cannot produce one of its own. It is, therefore, a defective virus and cannot replicate without the help of hepatitis B virus. In similar manner, the virophage 'sputnik' is dependent on mimivirus, which infects the protozoan *Acanthamoeba castellanii*. These viruses that are dependent on the presence of other virus species in the host cell are called *satellites* and may represent evolutionary intermediates of viroids and viruses. In the past, there were problems with all of these hypotheses: the regressive hypothesis did not explain why even the smallest of cellular parasites do not resemble viruses in any way. The escape hypothesis did not explain the complex capsids and other structures on virus particles. The virus-first hypothesis contravened the definition of viruses in that they require host cells. Viruses are now recognised as ancient and to have origins that predate the divergence of life into the three domains. This discovery has led modern virologists to reconsider and re-evaluate these three classical hypotheses.

The evidence for an ancestral world of RNA cells and computer analysis of viral and host DNA sequences are giving a better understanding of the evolutionary relationships between different viruses and may help identify the ancestors of modern viruses. To date, such analyses have not proved which of these hypotheses is correct. However, it seems unlikely that all currently known viruses have a common ancestor, and viruses have probably arisen numerous times in the past by one or more mechanisms.

Prions are infectious protein molecules that do not contain DNA or RNA. They cause an infection in sheep called scrapie and cattle bovine spongiform encephalopathy ("mad cow" disease). In humans they cause kuru and Creutzfeldt-Jakob disease. They are able to replicate because some proteins can exist in two different shapes and the prion changes the normal shape of a host protein into the prion shape.

This starts a chain reaction where each prion protein converts many host proteins into more prions, and these new prions then go on to convert even more protein into prions. Although they are fundamentally different from viruses and viroids, their discovery gives credence to the idea that viruses could have evolved from self-replicating molecules.

Microbiology

Life Properties

Opinions differ on whether viruses are a form of life, or organic structures that interact with living organisms. They have been described as "organisms at the edge of life", since they resemble organisms in that they possess genes and evolve by natural selection, and reproduce by creating multiple copies of themselves through self-assembly.

Although they have genes, they do not have a cellular structure, which is often seen as the basic unit of life. Viruses do not have their own metabolism, and require a host cell to make new products. They therefore cannot naturally reproduce outside a host cell – although bacterial species such as rickettsia and chlamydia are considered living organisms despite the same limitation. Accepted forms of life use cell division to reproduce, whereas viruses spontaneously assemble within cells. They differ from autonomous growth of crystals as they inherit genetic mutations while being subject to natural selection. Virus self-assembly within host cells has implications for the study of the origin of life, as it lends further credence to the hypothesis that life could have started as self-assembling organic molecules.

Structure

Viruses display a wide diversity of shapes and sizes, called *morphologies*. Generally viruses are much smaller than bacteria. Most viruses that have been studied have a diameter between 20 and 300 nanometres. Some filoviruses have a total length of up to 1400 nm; their diameters are only about 80 nm.

Most viruses cannot be seen with a light microscope so scanning and transmission electron microscopes are used to visualize virions. To increase the contrast between viruses and the background, electron-dense "stains" are used. These are solutions of salts of heavy metals, such as tungsten, that scatter the electrons from regions covered with the stain. When virions are coated with stain (positive staining), fine detail is obscured. Negative staining overcomes this problem by staining the background only.

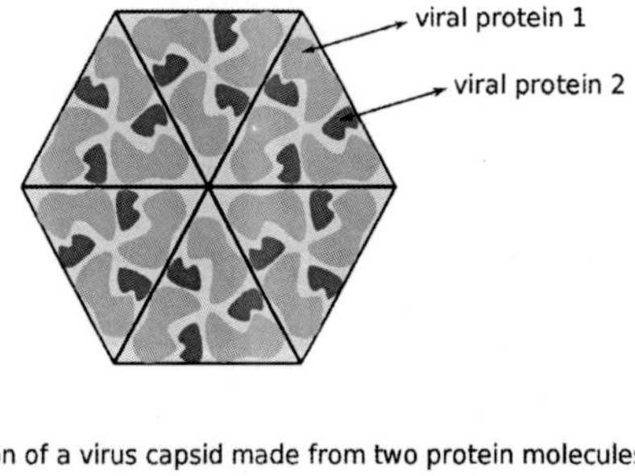

Figure : *Diagram of how a virus capsid can be constructed using multiple copies of just two protein molecules*

A complete virus particle, known as a virion, consists of nucleic acid surrounded by a protective coat of protein called a capsid. These are formed from identical protein subunits called capsomeres. Viruses can have a lipid "envelope" derived from the host cell membrane. The capsid is made from proteins encoded by the viral genome and its shape serves as the basis for morphological distinction. Virally coded protein subunits will self-assemble to form a capsid, generally requiring the presence of the virus genome.

Complex viruses code for proteins that assist in the construction of their capsid. Proteins associated with nucleic acid are known as nucleoproteins, and the association of viral capsid proteins with viral nucleic acid is called a nucleocapsid. The capsid and entire virus structure can be mechanically (physically) probed through atomic force microscopy.

In general, there are four main morphological virus types:

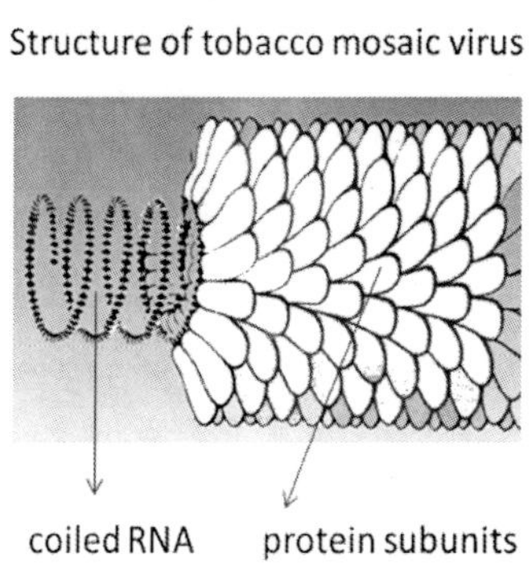

Figure: *RNA coiled in a helix of repeating protein subunits*

Helical: These viruses are composed of a single type of capsomer stacked around a central axis to form a helical structure, which may have a central cavity, or hollow tube.

This arrangement results in rod-shaped or filamentous virions: These can be short and highly rigid, or long and very flexible. The genetic material, in general, single-stranded RNA, but ssDNA in some cases, is bound into the protein helix by interactions between the negatively charged nucleic acid and positive charges on the protein. Overall, the length of a helical capsid is related to the length of the nucleic acid contained within it and the diameter is dependent on the size and arrangement of capsomers. The well-studied tobacco mosaic virus is an example of a helical virus.

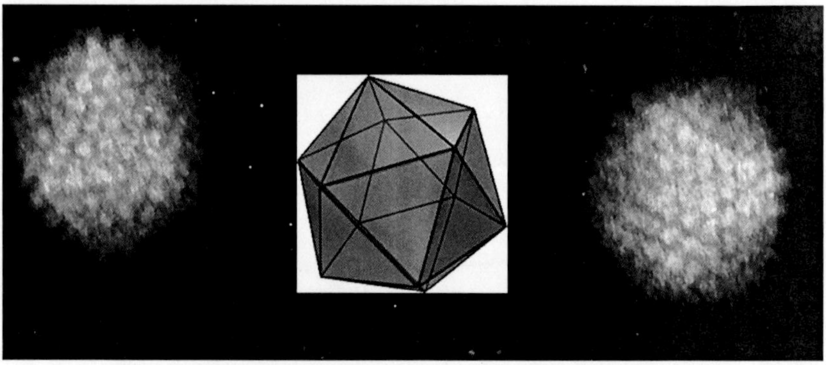

Figure: *Electron micrograph of icosahedral adenovirus*

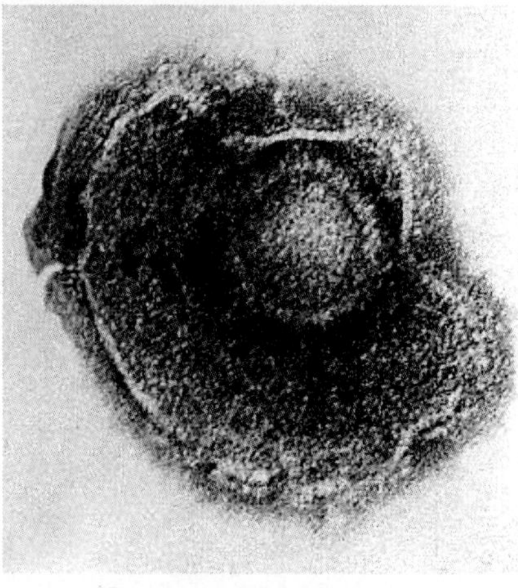

Figure : *Herpes viruses have a lipid envelope*

Icosahedral: Most animal viruses are icosahedral or near-spherical with icosahedral symmetry. A regular icosahedron is the optimum way of forming a closed shell from identical subunits. The minimum number of identical capsomers required is twelve, each composed of five identical subunits.

Many viruses, such as rotavirus, have more than twelve capsomers and appear spherical but they retain this symmetry. Capsomers at the apices are surrounded by five other capsomers and are called pentons. Capsomers on the triangular faces are surrounded by six others and are called hexons. Hexons are essentially flat and pentons, which form the 12 vertices, are curved. The same protein may act as the subunit of both the pentamers and hexamers or they may be composed of different proteins.

Although the isosahedral structure is extremely common among viruses because of size differences slight variations exist between virions. Given an asymmetric subunit on a triangular face of a regular isosahedron, with three subunits per face 60 such subunits can be placed in an equivalent manner. Most virions, because of their size, have more than 60 subunits. These variations have been classified on the basis of the quasi-equivalence principle proposed by Caspar and Klug. An isosahedral structure can be regarded as being constructed from 12 pentamers. The number of pentamers is fixed but the number of hexamers can vary. These shells can be constructed from pentamers and hexamers by minimising the number T (triangulation number) of nonequivalent locations that subunits occupy, with the T-number adopting the particular integer values 1, 3, 4, 7, 12, 13,...($T = h^2 + k^2 + hk$, with h, k equal to nonnegative integers). These shells always contain 12 pentamers plus 10 (T-1) hexamers. Although this classification can be applied to the majority of known viruses exceptions are known including the retroviruses where point mutations disrupt the symmetry.

Prolate: This is an isosahedron elongated along the fivefold axis and is a common arrangement of the heads of bacteriophages. Such a structure is composed of a cylinder with a cap at either end. The cylinder is composed of 10 triangles. The Q number, which can be can be any positive integer, specifies the number of triangles, composed of asymmetric subunits, that make up the 10 triangles of the cylinder. The caps are classified by the T number.

Envelope: Some species of virus envelop themselves in a modified form of one of the cell membranes, either the outer membrane surrounding an infected host cell or internal membranes such as nuclear membrane or endoplasmic reticulum, thus gaining an outer lipid bilayer known as a viral envelope. This membrane is studded with proteins coded for by the viral genome and host genome; the lipid membrane itself and any carbohydrates present originate entirely from the host. The influenza virus and HIV use this strategy. Most enveloped viruses are dependent on the envelope for their infectivity.

Complex: These viruses possess a capsid that is neither purely helical nor purely icosahedral, and that may possess extra structures such as protein tails or a complex outer wall. Some bacteriophages, such as Enterobacteria phage T4, have a complex structure consisting of an icosahedral head bound to a helical tail, which may have a hexagonal base plate with protruding protein tail fibres.

This tail structure acts like a molecular syringe, attaching to the bacterial host and then injecting the viral genome into the cell. The poxviruses are large, complex viruses that have an unusual morphology. The viral genome is associated with proteins within a central disk structure known as a nucleoid. The nucleoid is surrounded by a membrane and two lateral bodies of unknown function.

The virus has an outer envelope with a thick layer of protein studded over its surface. The whole virion is slightly pleiomorphic, ranging from ovoid to brick shape. Mimivirus is the largest characterised virus, with a capsid diameter of 400 nm. Protein filaments measuring 100 nm project from the surface. The capsid appears hexagonal under an electron microscope, therefore the capsid is probably icosahedral. In 2011, researchers discovered a larger virus on ocean floor of the coast of Las Cruces, Chile. Provisionally named *Megavirus chilensis*, it can be seen with a basic light microscope.

Some viruses that infect Archaea have complex structures that are unrelated to any other form of virus, with a wide variety of unusual shapes, ranging from spindle-shaped structures, to viruses that resemble hooked rods, teardrops or even bottles. Other archaeal viruses resemble the tailed bacteriophages, and can have multiple tail structures.

Genome

An enormous variety of genomic structures can be seen among viral species; as a group they contain more structural genomic diversity than plants, animals, archaea, or bacteria. There are millions of different types of viruses, although only about 5,000 of them have been described in detail. A virus has either DNA or RNA genes and is called a DNA virus or a RNA virus respectively. The vast majority of viruses have RNA genomes. Plant viruses tend to have single-stranded RNA genomes and bacteriophages tend to have double-stranded DNA genomes.

Viral genomes are circular, as in the polyomaviruses, or linear, as in the adenoviruses. The type of nucleic acid is irrelevant to the shape of the genome. Among RNA viruses and certain DNA viruses, the genome is often divided up into separate parts, in which case it is called *segmented*. For RNA viruses, each segment often codes for only one protein and they are usually found together in one capsid. However, all segments are not required to be in the same virion for the virus to be infectious, as demonstrated by brome mosaic virus and several other plant viruses. A viral genome, irrespective of nucleic acid type, is almost always either single-stranded or double-stranded. Single-stranded genomes consist of an unpaired nucleic acid, analogous to one-half of a ladder split down the middle.

Double-stranded genomes consist of two complementary paired nucleic acids, analogous to a ladder. The virus particles of some virus families, such as those belonging to the *Hepadnaviridae*, contain a genome that is partially double-stranded and partially single-stranded.

For most viruses with RNA genomes and some with single-stranded DNA genomes, the single strands are said to be either positive-sense (called the plus-strand) or negative-sense (called the minus-strand), depending on whether or not they are complementary to the viral messenger RNA (mRNA).

Positive-sense viral RNA is in the same sense as viral mRNA and thus at least a part of it can be immediately translated by the host cell. Negative-sense viral RNA is complementary to mRNA and thus must be converted to positive-sense RNA by an RNA-dependent RNA polymerase before translation. DNA nomenclature for viruses with single-sense genomic ssDNA is similar to RNA nomenclature, in that the *coding strand* for the viral mRNA is complementary to it ("), and the *non-coding strand* is a copy of it (+).

However, several types of ssDNA and ssRNA viruses have genomes that are ambisense in that transcription can occur off both strands in a double-stranded replicative intermediate. Examples include geminiviruses, which are ssDNA plant viruses and arenaviruses, which are ssRNA viruses of animals.

Genome size varies greatly between species. The smallest viral genomes – the ssDNA circoviruses, family *Circoviridae* – code for only two proteins and have a genome size of only 2 kilobases; the largest – mimiviruses – have genome sizes of over 1.2 megabases and code for over one thousand proteins. RNA viruses generally have smaller genome sizes than DNA viruses because of a higher error-rate when replicating, and have a maximum upper size limit. Beyond this limit, errors in the genome when replicating render the virus useless or uncompetitive.

To compensate for this, RNA viruses often have segmented genomes – the genome is split into smaller molecules – thus reducing the chance that an error in a single-component genome will incapacitate the entire genome. In contrast, DNA viruses generally have larger genomes because of the high fidelity of their replication enzymes.

Single-strand DNA viruses are an exception to this rule, however, as mutation rates for these genomes can approach the extreme of the ssRNA virus case.

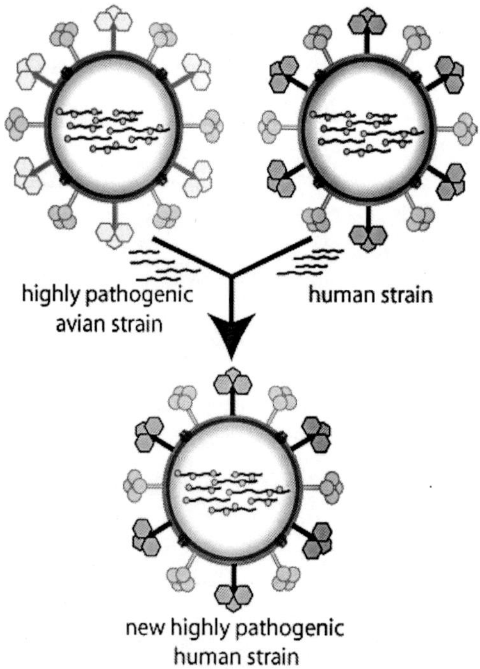

Figure: *How antigenic shift, or reassortment, can result in novel and highly pathogenic strains of human influenza*

Viruses undergo genetic change by several mechanisms. These include a process called genetic drift where individual bases in the DNA or RNA mutate to other bases. Most of these point mutations are "silent" – they do not change the protein that the gene encodes – but others can confer evolutionary advantages such as resistance to antiviral drugs.

Antigenic shift occurs when there is a major change in the genome of the virus. This can be a result of recombination or reassortment. When this happens with influenza viruses, pandemics might result. RNA viruses often exist as quasispecies or swarms of viruses of the same species but with slightly different genome nucleoside sequences. Such quasispecies are a prime target for natural selection.

Segmented genomes confer evolutionary advantages; different strains of a virus with a segmented genome can shuffle and combine genes and produce progeny viruses or (offspring) that have unique characteristics. This is called reassortment or *viral sex*. Genetic recombination is the process by which a strand of DNA is broken and then joined to the end of a different DNA molecule. This can occur when viruses infect cells simultaneously and studies of viral evolution have shown that recombination has been rampant in the species studied. Recombination is common to both RNA and DNA viruses.

Replication Cycle

Viral populations do not grow through cell division, because they are acellular. Instead, they use the machinery and metabolism of a host cell to produce multiple copies of themselves, and they *assemble* in the cell.

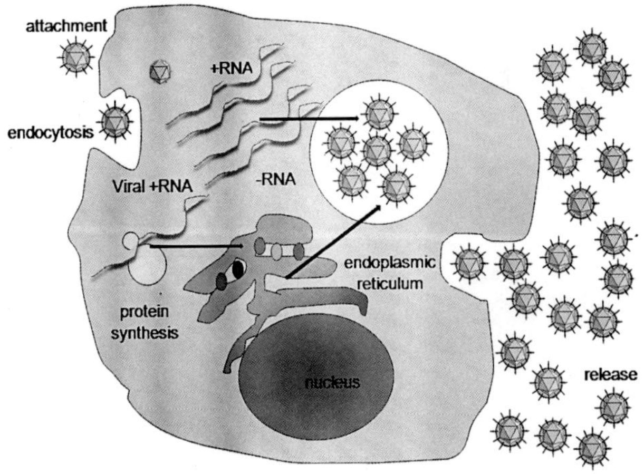

Figure: *A typical virus replication cycle*

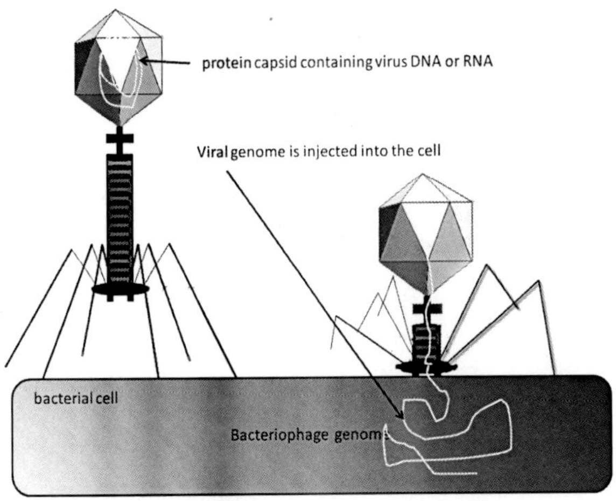

Figure: *Some bacteriophages inject their genomes into bacterial cells*

The life cycle of viruses differs greatly between species but there are six *basic* stages in the life cycle of viruses:

- *Attachment* is a specific binding between viral capsid proteins and specific receptors on the host cellular surface. This specificity determines the host range of a virus. For example, HIV infects a limited range of human leucocytes. This is because its surface protein, gp120, specifically interacts with the CD4 molecule – a chemokine receptor – which is most commonly found on the surface of CD4+ T-Cells. This mechanism has evolved to favour those viruses that infect only cells in which they are capable of replication. Attachment to the receptor can induce the viral envelope protein to undergo changes that results in the fusion of viral and cellular membranes, or changes of non-enveloped virus surface proteins that allow the virus to enter.

- *Penetration* follows attachment: Virions enter the host cell through receptor-mediated endocytosis or membrane fusion. This is often called viral entry. The infection of plant and, it is presumed, fungal cells is different from that of animal cells. Plants have a rigid cell wall made of cellulose, and fungi one of chitin, so most viruses can get inside these cells only after trauma to the cell wall. However, nearly all plant viruses (such as tobacco mosaic virus) can also move directly from cell to cell, in the form of single-stranded nucleoprotein complexes, through pores called plasmodesmata. This process requires movement proteins, which are virus-encoded proteins probably originally derived from plant proteins, which interact with the plasmodesmatal transport machinery Bacteria, like plants, have strong cell walls that a virus must breach to infect the cell. However, given that bacterial cell walls are much less thick than plant cell walls due to their much smaller size, some viruses have evolved mechanisms that inject their genome into the bacterial cell across the cell wall, while the viral capsid remains outside.
- *Uncoating* is a process in which the viral capsid is removed: This may be by degradation by viral enzymes or host enzymes or by simple dissociation; the end-result is the releasing of the viral genomic nucleic acid.
- *Replication* of viruses primarily involves multiplication of the genome. Replication involves synthesis of viral messenger RNA (mRNA) from "early" genes (with exceptions for positive sense RNA viruses), viral protein synthesis, possible assembly of viral proteins, then viral genome replication mediated by early or regulatory protein expression. This may be followed, for complex viruses with larger genomes, by one or more further rounds of mRNA synthesis: "late" gene expression is, in general, of structural or virion proteins.
- Following the structure-mediated self-*assembly* of the virus particles, some modification of the proteins often occurs. In viruses such as HIV, this modification (sometimes called maturation) occurs *after* the virus has been released from the host cell.

- Viruses can be *released* from the host cell by lysis, a process that kills the cell by bursting its membrane and cell wall if present: This is a feature of many bacterial and some animal viruses. Some viruses undergo a lysogenic cycle where the viral genome is incorporated by genetic recombination into a specific place in the host's chromosome. The viral genome is then known as a "provirus" or, in the case of bacteriophages a "prophage". Whenever the host divides, the viral genome is also replicated. The viral genome is mostly silent within the host; however, at some point, the provirus or prophage may give rise to active virus, which may lyse the host cells. Enveloped viruses (e.g., HIV) typically are released from the host cell by budding. During this process the virus acquires its envelope, which is a modified piece of the host's plasma or other, internal membrane.

The genetic material within virus particles, and the method by which the material is replicated, varies considerably between different types of viruses.

DNA viruses: The genome replication of most DNA viruses takes place in the cell's nucleus. If the cell has the appropriate receptor on its surface, these viruses enter the cell sometimes by direct fusion with the cell membrane (e.g. herpesviruses) or – more usually – by receptor-mediated endocytosis. Most DNA viruses are entirely dependent on the host cell's DNA and RNA synthesising machinery, and RNA processing machinery; however, viruses with larger genomes may encode much of this machinery themselves. In eukaryotes the viral genome must cross the cell's nuclear membrane to access this machinery, while in bacteria it need only enter the cell.

RNA viruses: Replication usually takes place in the cytoplasm. RNA viruses can be placed into four different groups depending on their modes of replication. The polarity (whether or not it can be used directly by ribosomes to make proteins) of single-stranded RNA viruses largely determines the replicative mechanism; the other major criterion is whether the genetic material is single-stranded or double-stranded. All RNA viruses use their own RNA replicase enzymes to create copies of their genomes.

Reverse transcribing viruses: These have ssRNA (*Retroviridae, Metaviridae, Pseudoviridae*) or dsDNA (*Caulimoviridae*, and *Hepadnaviridae*) in their particles. Reverse transcribing viruses with RNA genomes (retroviruses), use a DNA intermediate to replicate, whereas those with DNA genomes (pararetroviruses) use an RNA intermediate during genome replication. Both types use a reverse transcriptase, or RNA-dependent DNA polymerase enzyme, to carry out the nucleic acid conversion. Retroviruses integrate the DNA produced by reverse transcription into the host genome as a provirus as a part of the replication process; pararetroviruses do not, although integrated genome copies of especially plant pararetroviruses can give rise to infectious virus. They are susceptible to antiviral drugs that inhibit the reverse transcriptase enzyme, e.g. zidovudine and lamivudine. An example of the first type is HIV, which is a retrovirus. Examples of the second type are the *Hepadnaviridae*, which includes Hepatitis B virus.

Effects on the Host Cell

The range of structural and biochemical effects that viruses have on the host cell is extensive. These are called *cytopathic effects*. Most virus infections eventually result in the death of the host cell. The causes of death include cell lysis, alterations to the cell's surface membrane and apoptosis. Often cell death is caused by cessation of its normal activities because of suppression by virus-specific proteins, not all of which are components of the virus particle.

Some viruses cause no apparent changes to the infected cell. Cells in which the virus is latent and inactive show few signs of infection and often function normally. This causes persistent infections and the virus is often dormant for many months or years. This is often the case with herpes viruses. Some viruses, such as Epstein-Barr virus, can cause cells to proliferate without causing malignancy, while others, such as papillomaviruses, are established causes of cancer.

Host Range

Viruses are by far the most abundant parasites on earth, and they have been found to infect all types of cellular life including animals, plants, bacteria and fungi. However, different types of viruses can infect only a limited range of hosts and many are species-specific. Some, such as smallpox virus for example, can infect only one species – in this case humans, and are said to have a narrow host range. Other viruses, such as rabies virus, can infect different species of mammals and are said to have a broad range. The viruses that infect plants are harmless to animals, and most viruses that infect other animals are harmless to humans. The host range of some bacteriophages is limited to a single strain of bacteria and they can be used to trace the source of outbreaks of infections by a method called phage typing.

Classification

Classification seeks to describe the diversity of viruses by naming and grouping them on the basis of similarities. In 1962, André Lwoff, Robert Horne, and Paul Tournier were the first to develop a means of virus classification, based on the Linnaean hierarchical system. This system bases classification on phylum, class, order, family, genus, and species. Viruses were grouped according to their shared properties (not those of their hosts) and the type of nucleic acid forming their genomes. Later the International Committee on

Taxonomy of Viruses was formed. However, viruses are not classified on the basis of phylum or class, as their small genome size and high rate of mutation makes it difficult to determine their ancestry beyond Order. As such, the Baltimore Classification is used to supplement the more traditional hierarchy.

ICTV Classification

The International Committee on Taxonomy of Viruses (ICTV) developed the current classification system and wrote guidelines that put a greater weight on certain virus properties to maintain family uniformity. A unified taxonomy (a universal system for classifying viruses) has been established. The 7th lCTV Report formalized for the first time the concept of the virus species as the lowest taxon (group) in a branching hierarchy of viral taxa. However, at present only a small part of the total diversity of viruses has been studied, with analyses of samples from humans finding that about 20% of the virus sequences recovered have not been seen before, and samples from the environment, such as from seawater and ocean sediments, finding that the large majority of sequences are completely novel.

The general taxonomic structure is as follows:

- Order (-virales)
- Family (-viridae)
- Subfamily (-virinae)
- Genus (*-virus*)
- Species (*-virus*).

In the current (2010) ICTV taxonomy, six orders have been established, the Caudovirales, Herpesvirales, Mononegavirales, Nidovirales, Picornavirales and Tymovirales. A seventh order Ligamenvirales has also been proposed. The committee does not formally distinguish between subspecies, strains, and isolates. In total there are 6 orders, 87 families, 19 subfamilies, 348 genera, 2,885 species and about 3,000 types yet unclassified.

Baltimore Classification

The Nobel Prize-winning biologist David Baltimore devised the Baltimore classification system. The ICTV classification system is used in conjunction with the Baltimore classification system in modern virus classification. The Baltimore classification of viruses is based on the mechanism of mRNA production. Viruses must generate mRNAs from their genomes to produce proteins and replicate themselves, but different mechanisms are used to achieve this in each virus family. Viral genomes may be single-stranded (ss) or double-stranded (ds), RNA or DNA, and may or may not use reverse transcriptase (RT). Additionally, ssRNA viruses may be either sense (+) or antisense ("). This classification places viruses into seven groups:

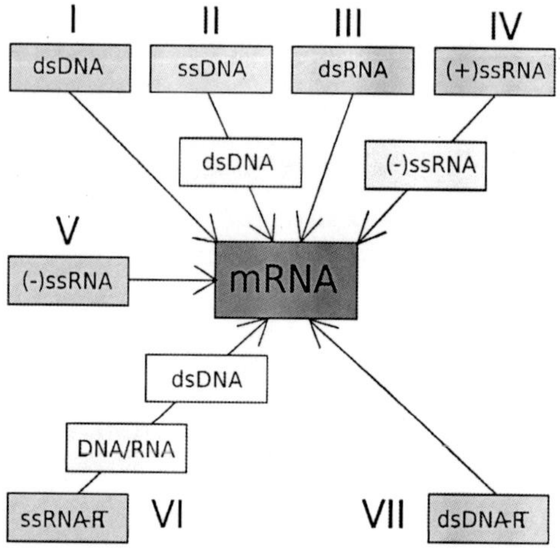

Figure: *The Baltimore Classification of viruses is based on the method of viral mRNA synthesis.*

- I: dsDNA viruses (e.g. Adenoviruses, Herpesviruses, Poxviruses)
- II: ssDNA viruses (+)sense DNA (e.g. Parvoviruses)
- III: dsRNA viruses (e.g. Reoviruses)
- IV: (+)ssRNA viruses (+)sense RNA (e.g. Picornaviruses, Togaviruses)
- V: (")ssRNA viruses (—)sense RNA (e.g. Orthomyxoviruses, Rhabdoviruses)
- VI: ssRNA-RT viruses (+)sense RNA with DNA intermediate in life-cycle (e.g. Retroviruses)
- VII: dsDNA-RT viruses (e.g. Hepadnaviruses).

Role in Human Disease

As an example of viral classification, the chicken pox virus, varicella zoster (VZV), belongs to the order Herpesvirales, family *Herpesviridae*, subfamily *Alphaherpesvirinae*, and genus *Varicellovirus*. VZV is in Group I of the Baltimore Classification because it is a dsDNA virus that does not use reverse transcriptase.

Examples of common human diseases caused by viruses include the common cold, influenza, chickenpox and cold sores. Many serious diseases such as ebola, AIDS, avian influenza and SARS are caused by viruses. The relative ability of viruses to cause disease is described in terms of virulence. Other diseases are under investigation as to whether they too have a virus as the causative agent, such as the possible connection between human herpes virus six (HHV6) and neurological diseases such as multiple sclerosis and chronic fatigue syndrome. There is controversy over whether the borna virus, previously thought to cause neurological diseases in horses, could be responsible for psychiatric illnesses in humans.

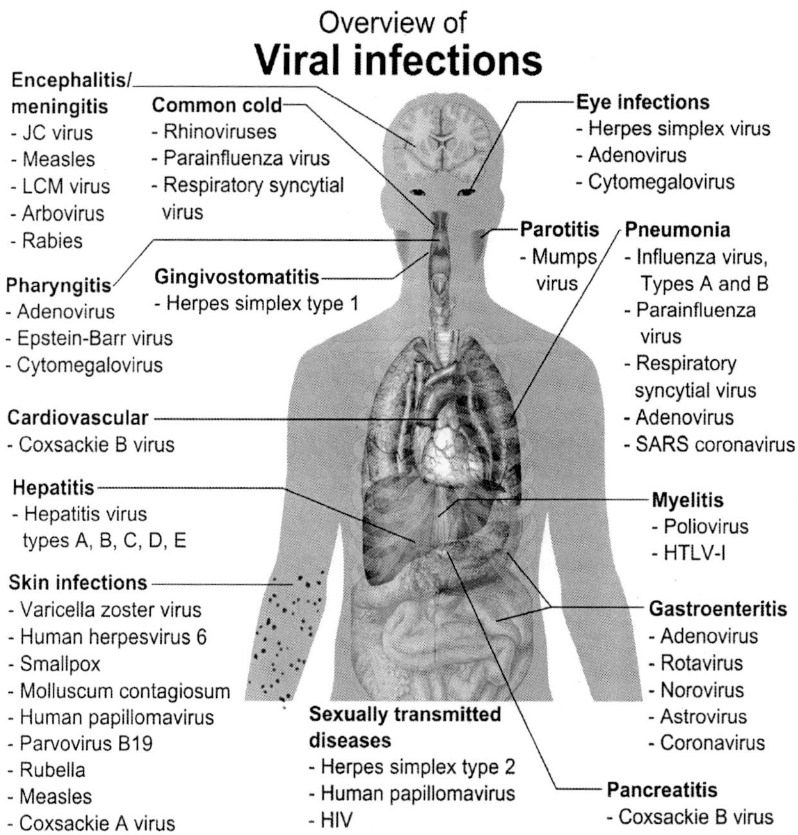

Figure: *Overview of the main types of viral infection and the most notable species involved.*

Viruses have different mechanisms by which they produce disease in an organism, which largely depends on the viral species. Mechanisms at the cellular level primarily include cell lysis, the breaking open and subsequent death of the cell. In multicellular organisms, if enough cells die, the whole organism will start to suffer the effects. Although viruses cause disruption of healthy homeostasis, resulting in disease, they may exist relatively harmlessly within an organism. An example would include the ability of the herpes simplex virus, which causes cold sores, to remain in a dormant state within the human body. This is called latency and is a characteristic of the herpes viruses including Epstein-Barr virus, which causes glandular fever, and varicella zoster virus, which causes chickenpox and shingles. Most people have been infected with at least one of these types of herpes virus. However, these latent viruses might sometimes be beneficial, as the presence of the virus can increase immunity against bacterial pathogens, such as *Yersinia pestis*. Some viruses can cause lifelong or chronic infections, where the viruses continue to replicate in the body despite the host's defence mechanisms. This is common in hepatitis B virus and hepatitis C virus infections. People chronically infected are known as carriers, as they serve as reservoirs of infectious virus. In populations with a high proportion of carriers, the disease is said to be endemic.

Epidemiology

Viral epidemiology is the branch of medical science that deals with the transmission and control of virus infections in humans. Transmission of viruses can be vertical, that is from mother to child, or horizontal, which means from person to person. Examples of vertical transmission include hepatitis B virus and HIV where the baby is born already infected with the virus. Another, more rare, example is the varicella zoster virus, which, although causing relatively mild infections in humans, can be fatal to the foetus and newborn baby.

Horizontal transmission is the most common mechanism of spread of viruses in populations. Transmission can occur when: body fluids are exchanged during sexual activity, e.g., HIV; blood is exchanged by contaminated transfusion or needle sharing, e.g., hepatitis C; a child is born to an infected mother, e.g., hepatitis B; exchange of saliva by mouth, e.g., Epstein-Barr virus; contaminated food or water is ingested, e.g., norovirus; aerosols containing virions are inhaled, e.g., influenza virus; and insect vectors such as mosquitoes penetrate the skin of a host, e.g., dengue.

The rate or speed of transmission of viral infections depends on factors that include population density, the number of susceptible individuals, (i.e., those not immune), the quality of healthcare and the weather. Epidemiology is used to break the chain of infection in populations during outbreaks of viral diseases. Control measures are used that are based on knowledge of how the virus is transmitted. It is important to find the source, or sources, of the outbreak and to identify the virus. Once the virus has been identified, the chain of transmission can sometimes be broken by vaccines. When vaccines are not available sanitation and disinfection can be effective. Often infected people are isolated from the rest of the community and those that have been exposed to the virus placed in quarantine. To control the outbreak of foot and mouth disease in cattle in Britain in 2001, thousands of cattle were slaughtered.

Most viral infections of humans and other animals have incubation periods during which the infection causes no signs or symptoms. Incubation periods for viral diseases range from a few days to weeks but are known for most infections. Somewhat overlapping, but mainly following the incubation period, there is a period of communicability; a time when an infected individual or animal is contagious and can infect another person or animal. This too is known for many viral infections and knowledge the length of both periods is important in the control of outbreaks. When outbreaks cause an unusually high proportion of cases in a population, community or region they are called epidemics. If outbreaks spread worldwide they are called pandemics.

Epidemics and Pandemics

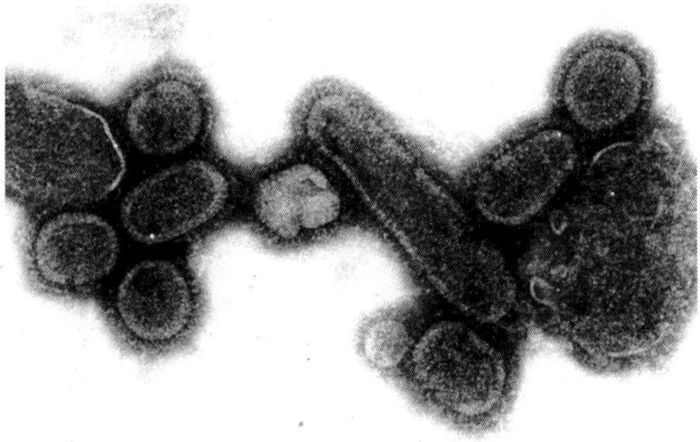

Figure: *Transmission electron microscope image of a recreated 1918 influenza virus*

Native American populations were devastated by contagious diseases, in particular, smallpox, brought to the Americas by European colonists. It is unclear how many Native Americans were killed by foreign diseases after the arrival of Columbus in the Americas, but the numbers have been estimated to be close to 70% of the indigenous population. The damage done by this disease significantly aided European attempts to displace and conquer the native population.

A pandemic is a worldwide epidemic. The 1918 flu pandemic, commonly referred to as the Spanish flu, was a category 5 influenza pandemic caused by an unusually severe and deadly influenza A virus. The victims were often healthy young adults, in contrast to most influenza outbreaks, which predominantly affect juvenile, elderly, or otherwise-weakened patients.

The Spanish flu pandemic lasted from 1918 to 1919. Older estimates say it killed 40–50 million people, while more recent research suggests that it may have killed as many as 100 million people, or 5% of the world's population in 1918.

Most researchers believe that HIV originated in sub-Saharan Africa during the 20th century; it is now a pandemic, with an estimated 38.6 million people now living with the disease worldwide. The Joint United Nations Programme on HIV/AIDS (UNAIDS) and the World Health Organization (WHO) estimate that AIDS has killed more than 25 million people since it was first recognized on June 5, 1981, making it one of the most destructive epidemics in recorded history. In 2007 there were 2.7 million new HIV infections and 2 million HIV-related deaths.

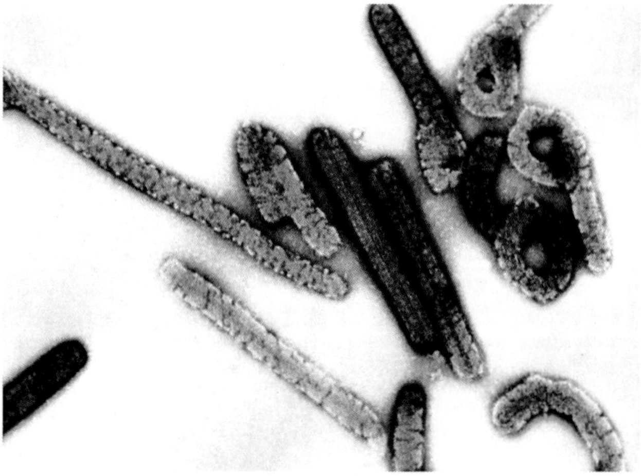

Figure: *Marburg virus*

Several highly lethal viral pathogens are members of the *Filoviridae*. Filoviruses are filament-like viruses that cause viral hemorrhagic fever, and include the ebola and marburg viruses. The Marburg virus attracted widespread press attention in April 2005 for an outbreak in Angola. Beginning in October 2004 and continuing into 2005, the outbreak was the world's worst epidemic of any kind of viral hemorrhagic fever.

Cancer

Viruses are an established cause of cancer in humans and other species. Viral cancers occur only in a minority of infected persons (or animals). Cancer viruses come from a range of virus families, including both RNA and DNA viruses, and so there is no single type of "oncovirus" (an obsolete term originally used for acutely transforming retroviruses).

The development of cancer is determined by a variety of factors such as host immunity and mutations in the host. Viruses accepted to cause human cancers include some genotypes of human papillomavirus, hepatitis B virus, hepatitis C virus, Epstein-Barr virus, Kaposi's sarcoma-associated herpesvirus and human T-lymphotropic virus. The most recently discovered human cancer virus is a polyomavirus (Merkel cell polyomavirus) that causes most cases of a rare form of skin cancer called Merkel cell carcinoma. Hepatitis viruses can develop into a chronic viral infection that leads to liver cancer. Infection by human T-lymphotropic virus can lead to tropical spastic paraparesis and adult T-cell leukemia. Human papillomaviruses are an established cause of cancers of cervix, skin, anus, and penis. Within the *Herpesviridae*, Kaposi's sarcoma-associated herpesvirus causes Kaposi's sarcoma and body cavity lymphoma, and Epstein–Barr virus causes Burkitt's lymphoma, Hodgkin's lymphoma, B lymphoproliferative disorder, and nasopharyngeal carcinoma. Merkel cell polyomavirus closely related to SV40 and mouse polyomaviruses that have been used as animal models for cancer viruses for over 50 years.

Host Defence Mechanisms

The body's first line of defence against viruses is the innate immune system. This comprises cells and other mechanisms that defend the host from infection in a non-specific manner. This means that the cells of the innate system recognise, and respond to, pathogens in a generic way, but, unlike the adaptive immune system, it does not confer long-lasting or protective immunity to the host.

RNA interference is an important innate defence against viruses. Many viruses have a replication strategy that involves double-stranded RNA (dsRNA). When such a virus infects a cell, it releases its RNA molecule or molecules, which immediately bind to a protein complex called dicer that cuts the RNA into smaller pieces. A biochemical pathway called the RISC complex is activated, which degrades the viral mRNA and the cell survives the infection. Rotaviruses avoid this mechanism by not uncoating fully inside the cell and by releasing newly produced mRNA through pores in the particle's inner capsid. The genomic dsRNA remains protected inside the core of the virion.

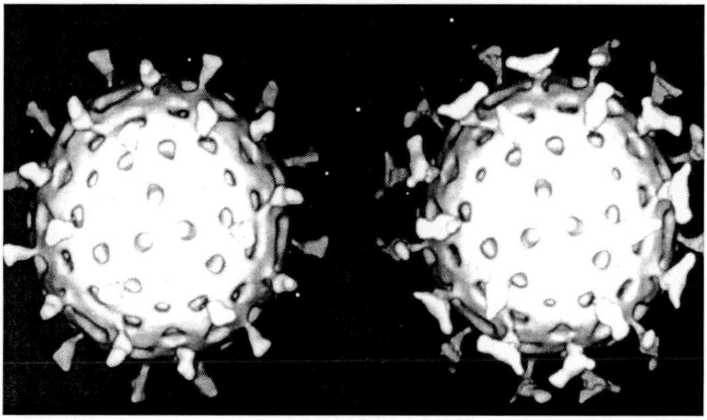

Figure: *Two rotaviruses: the one on the right is coated with antibodies that stop its attaching to cells and infecting them*

When the adaptive immune system of a vertebrate encounters a virus, it produces specific antibodies that bind to the virus and render it non-infectious. This is called humoral immunity. Two types of antibodies are important. The first, called IgM, is highly effective at neutralizing viruses but is produced by the cells of the immune system only for a few weeks. The second, called IgG, is produced indefinitely. The presence of IgM in the blood of the host is used to test for acute infection, whereas IgG indicates an infection sometime in the past. IgG antibody is measured when tests for immunity are carried out.

A second defence of vertebrates against viruses is called cell-mediated immunity and involves immune cells known as T cells. The body's cells constantly display short fragments of their proteins on the cell's surface, and, if a T cell recognises a suspicious viral fragment there, the host cell is destroyed by *killer T* cells and the virus-specific T-cells proliferate. Cells such as the macrophage are specialists at this antigen presentation. The production of interferon is an important host defence mechanism. This is a hormone produced by the body when viruses are present. Its role in immunity is complex; it eventually stops the viruses from reproducing by killing the infected cell and its close neighbours.

Not all virus infections produce a protective immune response in this way. HIV evades the immune system by constantly changing the amino acid sequence of the proteins on the surface of the virion. These persistent viruses evade immune control by sequestration, blockade of antigen presentation, cytokine resistance, evasion of natural killer cell activities, escape from apoptosis, and antigenic shift. Other viruses, called *neurotropic viruses*, are disseminated by neural spread where the immune system may be unable to reach them.

Prevention and Treatment

Because viruses use vital metabolic pathways within host cells to replicate, they are difficult to eliminate without using drugs that cause toxic effects to host cells in general. The most effective medical approaches to viral diseases are vaccinations to provide immunity to infection, and antiviral drugs that selectively interfere with viral replication.

Vaccines

Vaccination is a cheap and effective way of preventing infections by viruses. Vaccines were used to prevent viral infections long before the discovery of the actual viruses. Their use has resulted in a dramatic decline in morbidity (illness) and mortality (death) associated with viral infections such as polio, measles, mumps and rubella. Smallpox infections have been eradicated. Vaccines are available to prevent over thirteen viral infections of humans, and more are used to prevent viral infections of animals. Vaccines can consist of live-attenuated or killed viruses, or viral proteins (antigens). Live vaccines contain weakened forms of the virus, which do not cause the disease but, nonetheless, confer immunity. Such viruses are called attenuated. Live vaccines can be dangerous when given to people with a weak immunity, (who are described as immunocompromised), because in these people, the weakened virus can cause the original disease. Biotechnology and genetic engineering techniques are used to produce subunit vaccines.

These vaccines use only the capsid proteins of the virus. Hepatitis B vaccine is an example of this type of vaccine. Subunit vaccines are safe for immunocompromised patients because they cannot cause the disease. The yellow fever virus vaccine, a live-attenuated strain called 17D, is probably the safest and most effective vaccine ever generated.

Antiviral Drugs

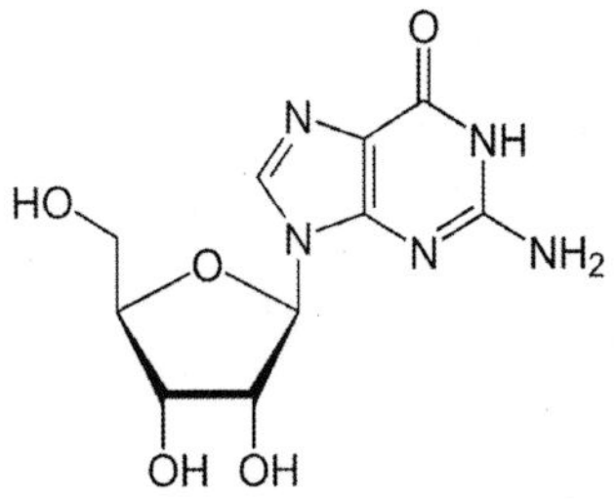

Figure: *Guanosine*

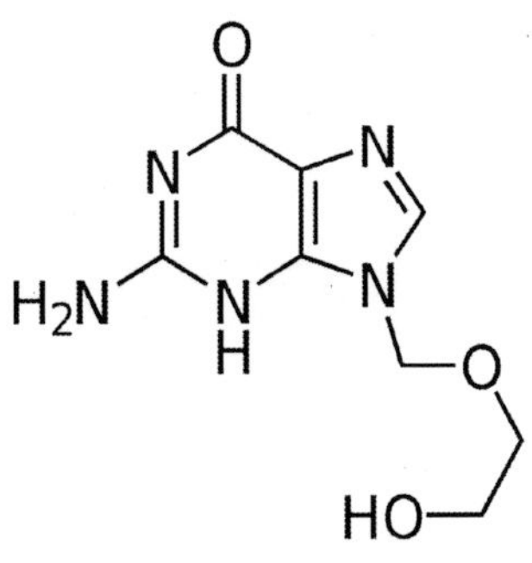

Figure: *The guanosine analogue Aciclovir*

Antiviral drugs are often nucleoside analogues, (fake DNA building-blocks), which viruses mistakenly incorporate into their genomes during replication. The life-cycle of the virus is then halted because the newly synthesised DNA is inactive. This is because these analogues lack the hydroxyl groups, which, along with phosphorus atoms, link together to form the strong "backbone" of the DNA molecule.

This is called DNA chain termination. Examples of nucleoside analogues are aciclovir for Herpes simplex virus infections and lamivudine for HIV and Hepatitis B virus infections. Aciclovir is one of the oldest and most frequently prescribed antiviral drugs. Other antiviral drugs in use target different stages of the viral life cycle. HIV is dependent on a proteolytic enzyme called the HIV-1 protease for it to become fully infectious. There is a large class of drugs called protease inhibitors that inactivate this enzyme. Hepatitis C is caused by an RNA virus. In 80% of people infected, the disease is chronic, and without treatment, they are infected for the remainder of their lives. However, there is now an effective treatment that uses the nucleoside analogue drug ribavirin combined with interferon. The treatment of chronic carriers of the hepatitis B virus by using a similar strategy using lamivudine has been developed.

Infection in Other Species

Plants

Figure: *Peppers infected by mild mottle virus*

Viruses infect all cellular life and, although viruses occur universally, each cellular species has its own specific range that often infect only that species. Some viruses, called satellites, can only replicate within cells that have already been infected by another virus. Viruses are important pathogens of livestock. Diseases such as Foot and Mouth Disease and bluetongue are caused by viruses. Companion animals such as cats, dogs, and horses, if not vaccinated, are susceptible to serious viral infections.

Canine parvovirus is caused by a small DNA virus and infections are often fatal in pups. Like all invertebrates, the honey bee is susceptible to many viral infections. Fortunately, most viruses co-exist harmlessly in their host and cause no signs or symptoms of disease. There are many types of plant virus, but often they cause only a loss of yield, and it is not economically viable to try to control them. Plant viruses are often spread from plant to plant by organisms, known as *vectors*.

These are normally insects, but some fungi, nematode worms, and single-celled organisms have been shown to be vectors. When control of plant virus infections is considered economical, for perennial fruits, for example, efforts are concentrated on killing the vectors and removing alternate hosts such as weeds. Plant viruses are harmless to humans and other animals because they can reproduce only in living plant cells.

Plants have elaborate and effective defence mechanisms against viruses. One of the most effective is the presence of so-called resistance (R) genes. Each R gene confers resistance to a particular virus by triggering localised areas of cell death around the infected cell, which can often be seen with the unaided eye as large spots.

This stops the infection from spreading. RNA interference is also an effective defence in plants. When they are infected, plants often produce natural disinfectants that kill viruses, such as salicylic acid, nitric oxide, and reactive oxygen molecules. Plant virus particles or virus-like particles (VLPs) have applications in both biotechnology and nanotechnology. The capsids of most plant viruses are simple and robust structures and can be produced in large quantities either by the infection of plants or by expression in a variety of heterologous systems. Plant virus particles can be modified genetically and chemically to encapsulate foreign material and can be incorporated into supramolecular structures for use in biotechnology.

Bacteria

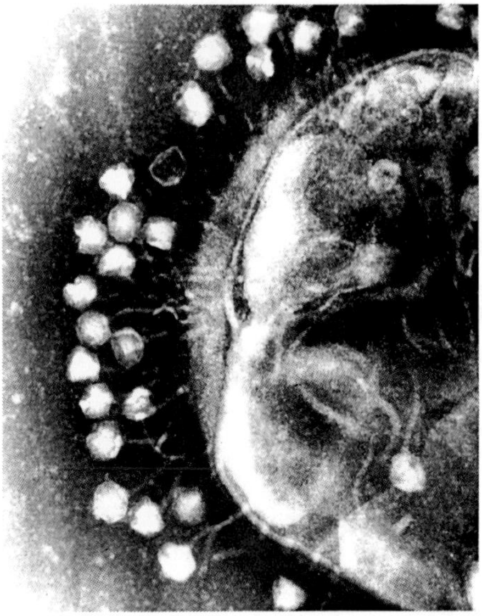

Figure: *Transmission electron micrograph of multiple bacteriophages attached to a bacterial cell wall*

Bacteriophages are a common and diverse group of viruses and are the most abundant form of biological entity in aquatic environments – there are up to ten times more of these viruses in the oceans than there are bacteria, reaching levels of 250,000,000 bacteriophages per millilitre of seawater. These viruses infect specific bacteria by binding to surface receptor molecules and then entering the cell.

Within a short amount of time, in some cases just minutes, bacterial polymerase starts translating viral mRNA into protein. These proteins go on to become either new virions within the cell, helper proteins, which help assembly of new virions, or proteins involved in cell lysis. Viral enzymes aid in the breakdown of the cell membrane, and, in the case of the T4 phage, in just over twenty minutes after injection over three hundred phages could be released.

The major way bacteria defend themselves from bacteriophages is by producing enzymes that destroy foreign DNA. These enzymes, called restriction endonucleases, cut up the viral DNA that bacteriophages inject into bacterial cells. Bacteria also contain a system that uses CRISPR sequences to retain fragments of the genomes of viruses that the bacteria have come into contact with in the past, which allows them to block the virus's replication through a form of RNA interference. This genetic system provides bacteria with acquired immunity to infection.

Archaea

Some viruses replicate within archaea: these are double-stranded DNA viruses with unusual and sometimes unique shapes. These viruses have been studied in most detail in the thermophilic archaea, particularly the orders Sulfolobales and Thermoproteales. Defences against these viruses may involve RNA interference from repetitive DNA sequences within archaean genomes that are related to the genes of the viruses.

Role in Aquatic Ecosystems

Viruses are the most abundant biological entity in aquatic environments: a teaspoon of seawater contains about one million of them. They are essential to the regulation of saltwater and freshwater ecosystems. Most of these viruses are bacteriophages, which are harmless to plants and animals. They infect and destroy the bacteria in aquatic microbial communities, comprising the most important mechanism of recycling carbon in the marine environment. The organic molecules released from the bacterial cells by the viruses stimulates fresh bacterial and algal growth.

Microorganisms constitute more than 90% of the biomass in the sea. It is estimated that viruses kill approximately 20% of this biomass each day and that there are 15 times as many viruses in the oceans as there are bacteria and archaea. Viruses are the main agents responsible for the rapid destruction of harmful algal blooms, which often kill other marine life. The number of viruses in the oceans decreases further offshore and deeper into the water, where there are fewer host organisms. The effects of marine viruses are far-reaching; by increasing the amount of photosynthesis in the oceans, viruses are indirectly responsible for reducing the amount of carbon dioxide in the atmosphere by approximately 3 gigatonnes of carbon per year.

Like any organism, marine mammals are susceptible to viral infections. In 1988 and 2002, thousands of harbour seals were killed in Europe by phocine distemper virus. Many other viruses, including caliciviruses, herpesviruses, adenoviruses and parvoviruses, circulate in marine mammal populations.

Role in Evolution

Viruses are an important natural means of transferring genes between different species, which increases genetic diversity and drives evolution. It is thought that viruses played a central role in the early evolution, before the diversification of bacteria, archaea and eukaryotes and at the time of the last universal common ancestor of life on Earth. Viruses are still one of the largest reservoirs of unexplored genetic diversity on Earth.

Applications

Life Sciences and Medicine

Viruses are important to the study of molecular and cellular biology as they provide simple systems that can be used to manipulate and investigate the functions of cells. The study and use of viruses have provided valuable information about aspects of cell biology. For example, viruses have been useful in the study of genetics and helped our understanding of the basic mechanisms of molecular genetics, such as DNA replication, transcription, RNA processing, translation, protein transport, and immunology.

Geneticists often use viruses as vectors to introduce genes into cells that they are studying. This is useful for making the cell produce a foreign substance, or to study the effect of introducing a new gene into the genome. In similar fashion, virotherapy uses viruses as vectors to treat various diseases, as they can specifically target cells and DNA. It shows promising use in the treatment of cancer and in gene therapy. Eastern European scientists have used phage therapy as an alternative to antibiotics for some time, and interest in this approach is increasing, because of the high level of antibiotic resistance now found in some pathogenic bacteria. Expression of heterologous proteins by viruses is the basis of several manufacturing processes that are currently being used for the production of various proteins such as vaccine antigens and antibodies. Industrial processes have been recently developed using viral vectors and a number of pharmaceutical proteins are currently in preclinical and clinical trials.

Materials Science and Nanotechnology

Current trends in nanotechnology promise to make much more versatile use of viruses. From the viewpoint of a materials scientist, viruses can be regarded as organic nanoparticles. Their surface carries specific tools designed to cross the barriers of their host cells.

The size and shape of viruses, and the number and nature of the functional groups on their surface, is precisely defined. As such, viruses are commonly used in materials science as scaffolds for covalently linked surface modifications. A particular quality of viruses is that they can be tailored by directed evolution. The powerful techniques developed by life sciences are becoming the basis of engineering approaches towards nanomaterials, opening a wide range of applications far beyond biology and medicine.

Because of their size, shape, and well-defined chemical structures, viruses have been used as templates for organising materials on the nanoscale. Recent examples include work at the Naval Research Laboratory in Washington, DC, using Cowpea Mosaic Virus (CPMV) particles to amplify signals in DNA microarray based sensors. In this application, the virus particles separate the fluorescent dyes used for signalling to prevent the formation of non-fluorescent dimers that act as quenchers. Another example is the use of CPMV as a nanoscale breadboard for molecular electronics.

Synthetic Viruses

Many viruses can be synthesized de novo ("from scratch") and the first synthetic virus was created in 2002. Although somewhat of a misconception, it is not the actual virus that is synthesized, but rather its DNA genome (in case of a DNA virus), or a cDNA copy of its genome (in case of RNA viruses). For many virus families the naked synthetic DNA or RNA (once enzymatically converted back from the synthetic cDNA) is infectious when introduced into a cell.

That is, they contain all the necessary information to produce new viruses. This technology is now being used to investigate novel vaccine strategies. The ability to synthesize viruses has far-reaching consequences, since viruses can no longer be regarded as extinct, as long as the information of their genome sequence is known and permissive cells are available. Currently, the full-length genome sequences of 2408 different viruses (including smallpox) are publicly available at an online database, maintained by the National Institute of Health.

Weapons

The ability of viruses to cause devastating epidemics in human societies has led to the concern that viruses could be weaponised for biological warfare. Further concern was raised by the successful recreation of the infamous 1918 influenza virus in a laboratory. The smallpox virus devastated numerous societies throughout history before its eradication.

There are officially only two centres in the world that keep stocks of smallpox virus – the Russian Vector laboratory, and the United States Centres for Disease Control. But fears that it may be used as a weapon are not totally unfounded; the vaccine for smallpox has sometimes severe side-effects – during the last years before the eradication of smallpox disease more people became seriously ill as a result of vaccination than did people from smallpox – and smallpox vaccination is no longer universally practised.

Thus, much of the modern human population has almost no established resistance to smallpox. A viral Antigen is an antigen with multiple antigenicities that is protein in nature, strain-specific, and closely associated with the virus particle. A viral antigen is a protein encoded by the viral genome. A viral protein is an antigen specified by the viral genome that can be detected by a specific immunological response.

Viruses are infectious pathogens that cause serious diseases & major threats for global public health, such as influenza, hepatitis, and AIDS. Virus is a sub-micrometer particle that has DNA or RNA packed in a shell called capsid. Viral antigens protrude from the capsid and often fulfill important function in docking to the host cell, fusion, and injection of viral DNA/RNA. Antibody-based immune responses form a first layer of protection of the host from viral infection; however, in many cases a vigorous cellular immune response mediated by T-cells and NK-cells is required for effective viral clearance.

When cellular immunity is unable to clear the virus, the infection can become chronic, and serum antibodies to the viral pathogen are used as first indicator for the diagnosis of the disease.

ELISAs provide a valuable tool in the detection and diagnosis of virus infection. The ability to produce recombinant viral proteins will ensure that future ELISAs are safe, specific and rapid. Even when a virus cannot be cultured, provided gene sequence is available, it is possible to rapidly respond to emerging viruses and new viral strains of existing pathogens.

Recombinant viral antigens contain part of viral sequence meaning that the recombinant antigen contains a region which can be recognized by different antibodies produced by different individuals. This reduces the risk of false negatives which can occur with synthetic peptides, which contain only a small portion of the entire protein. If an individual infected with a viral antigen makes antibodies to a part of the protein not included in the synthetic peptides, a false negative results.

Recombinant viral protein usually contains a fusion protein/partner which produces superior attachment to assay surfaces such as wells. For this reason, smaller amounts of recombinant protein will produce the same results as larger amounts of unfused protein. The choice of fusion partner prevents false positives, allowing superior adhesion without incorrect results.

Recombinant Viral proteins are expressed in bacteria, yeast, mammalian cells, and viruses. *E. Coli* cells were first to be used for this purpose but the expressed proteins were not glycosylated, which was a major drawback since many of the immunogenic proteins of viruses such as the envelope glycoproteins, were glycosylated. Nevertheless, in many instances, it was demonstrated that the non-glycosylated protein backbone was just as immunogenic. The obvious advantage of recombinant viral antigens is that they are available in unlimited quantities and the production and quality control processes is simple.

Advantages of defined using recombinant viral antigens:

1. Production and quality control is simple.
2. No nucleic acids or other viral or external proteins, therefore less toxic.
3. Safer in cases where viruses are oncogenic or establish a persistent infection.
4. Feasible even if virus cannot be cultivated.

Disadvantages:

1. May be less immunogenic than conventional inactivated whole-virus vaccines.
2. Requires adjuvant .
3. Fails to elicit CMI.

Facts about Viral Antigens:

1. A Viral Protein Mimics its Way into cells.
2. Viral Protein Helps Infected T Cells Stick To Uninfected Cells.
3. The Viral Protein A238L Inhibits Cyclooxygenase-2 Expression through a Nuclear Factor of Activated T Cell-dependent Trans-activation Pathway.
4. Viral Protein is an effective preventative against ear infection.
5. HIV-1 Viral Protein R Induces Apoptosis via a direct effect on the Mitochondrial Permeability Transition Pore.

6. The level of viral antigen presented by Hepatocytes Influences CD8 T-Cell Function.
7. Antigen-presenting cells from calves persistently infected with bovine viral diarrhea virus, a member of the flaviviridae, are not compromised in their ability to present viral antigen.
8. There is a difference in the distribution and spread of a viral antigen, development of lesions and correlation between presence of viral antigen and lesions.
9. The absence of viral antigens on the surface of equine herpesvirus-1-infected peripheral blood mononuclear cells is a strategy to avoid complement-mediated lysis.
10. Viral protein influences key Cell-signalling pathway.
11. A viral protein produced by cancer-causing virus influences a key signaling pathway in the immune cells that the virus infects. This stimulates the cells to divide.
12 Protection by recombinant viral proteins against a respiratory virulent avian metapneumovirus has been achieved.
13. Viral O-acetylesterases are found in influenza C viruses and Coronaviruses. Viral O-acetylesterases remove cellular receptors from the surface of target cells which destroys the receptor. Recombinant viral O-acetylesterases derived from Sf9 insect cells as chimeric proteins fused to eGFP specifically hydrolyze 9-O-acetylated sialic acids, while that of sialodacryoadenitis virus, a rat coronavirus related to mouse hepatitis virus, is specific for 4-O-acetylated sialic acid. The recombinant esterases were shown to specifically de-O-acetylate sialic acids on glycoconjugates. The recombinant viral proteins can be used to unambiguously identify O-acetylated acids.

An antigen is a foreign molecule that, when introduced into the body, triggers the production of an antibody by the immune system. The immune system will then kill or neutralize the antigen that is recognised as a foreign and potentially harmful invader.

These invaders can be molecules such as pollen or cells such as bacteria. The term originally came from antibody generator and was a molecule that binds specifically to an antibody, but the term now also refers to any molecule or molecular fragment that can be bound by a major histocompatibility complex (MHC) and presented to a T-cell receptor. "Self" antigens are usually tolerated by the immune system; whereas "Non-self" antigens are identified as invaders and attacked by the immune system.

An immunogen is a specific type of antigen. An immunogen is a substance that is able to provoke an adaptive immune response if injected on its own. An immunogen is able to induce an immune response, whereas an antigen is able to combine with the products of an immune response once they are made. The overlapping concepts of immunogenicity and antigenicity are, therefore, subtly different. According to a current textbook:

Immunogenicity is the ability to induce a humoral and/or cell-mediated immune response

Antigenicity is the ability to combine specifically with the final products of the immune response (i.e. secreted antibodies and/or surface receptors on T-cells). Although all molecules that have the property of immunogenicity also have the property of antigenicity, the reverse is not true." At the molecular level, an antigen is characterised by its ability to be "bound" at the antigen-binding site of an antibody. Note also that antibodies tend to discriminate between the specific molecular structures presented on the surface of the antigen. Antigens are usually proteins or polysaccharides.

This includes parts (coats, capsules, cell walls, flagella, fimbrae, and toxins) of bacteria, viruses, and other microorganisms. Lipids and nucleic acids are antigenic only when combined with proteins and polysaccharides. Non-microbial exogenous (non-self) antigens can include pollen, egg white, and proteins from transplanted tissues and organs or on the surface of transfused blood cells. Vaccines are examples of immunogenic antigens intentionally administered to induce acquired immunity in the recipient.

Cells present their immunogenic-antigens to the immune system via a histocompatibility molecule. Depending on the antigen presented and the type of the histocompatibility molecule, several types of immune cells can become activated.

Epitope

An epitope, also known as *antigenic determinant*, is the part of an antigen that is recognised by the immune system, specifically by antibodies, B cells, or T cells. The part of an antibody that recognises the epitope is called a paratope. Although epitopes are usually thought to be derived from non-self proteins, sequences derived from the host that can be recognised are also classified as epitopes.

The epitopes of protein antigens are divided into two categories, conformational epitopes and linear epitopes, based on their structure and interaction with the paratope. A conformational epitope is composed of discontinuous sections of the antigen's amino acid sequence. These epitopes interact with the paratope based on the 3-D surface features and shape or tertiary structure of the antigen. Most epitopes are conformational.

By contrast, linear epitopes interact with the paratope based on their primary structure. A linear epitope is formed by a continuous sequence of amino acids from the antigen.

The distinct molecular surface features of an antigen capable of being bound by an antibody (a.k.a. *antigenic determinant*). Antigenic molecules, normally being "large" biological polymers, usually present several surface features that can act as points of interaction for specific antibodies. Any such distinct molecular feature constitutes an epitope. Therefore, most antigens have the potential to be bound by several distinct antibodies, each of which is specific to a particular epitope. Using the "lock and key" metaphor, the antigen itself can be seen as a string of keys - any epitope being a "key" - each of which can match a different lock. Different antibody idiotypes, each having distinctly formed complementarity determining regions, correspond to the various "locks" that can match "the keys" (epitopes) presented on the antigen molecule.

Function

T-cell Epitopes

T-cell epitopes are presented on the surface of an antigen-presenting cell, where they are bound to MHC molecules. T-cell epitopes presented by MHC class I molecules are typically peptides between 8 and 11 amino acids in length, whereas MHC class II molecules present longer peptides, and non-classical MHC molecules also present non-peptidic epitopes such as glycolipids.

Cross-reactivity

Epitopes are sometimes cross-reactive. This property is exploited by the immune system in regulation by anti-idiotypic antibodies (originally proposed by Nobel laureate Niels Kaj Jerne). If an antibody binds to an antigen's epitope, the paratope could become the epitope for another antibody that will then bind to it. If this second antibody is of IgM class, its binding can upregulate the immune response; if the second antibody is of IgG class, its binding can downregulate the immune response.

Epitope Mapping

Epitopes can be mapped using protein microarrays, and with the ELISPOT or ELISA techniques.

Intensive research is currently taking place to design reliable tools that will predict epitopes on proteins.

Epitope mapping is the process of identifying the binding sites, or 'epitopes', of antibodies on their target antigens (which are proteins).

Identification and characterisation of the binding sites of antibodies can aid in the discovery and development of new therapeutics, vaccines, and diagnostics.

Epitopes (the binding sites on the protein) can be divided into linear and conformational. *Linear epitopes* are formed by a continuous sequence of amino acids in a protein, while *conformational epitopes* are composed of amino acids that are discontinuous in the protein sequence but are brought together upon three-dimensional protein folding. The vast majority of antigen-antibody interactions have conformational epitopes.

Methods for Epitope Mapping

Epitope mapping of complex target antigens, such as integral membrane proteins or multi-subunit proteins, is often challenging because of the difficulty in expressing and purifying these types of antigens.

There are several methods available for mapping antibody epitopes on target antigens :

- The gold standard approach is X-ray co-crystallography, which allows direct visualisation of the interaction between the antigen and antibody. However, this approach is technically challenging, requires large amounts of purified protein, and can be time-consuming and expensive.
- Array-based oligo-peptide scanning (sometimes called pepscan analysis): This technique uses a library of oligo-peptide sequences from overlapping and non-overlapping segments of a target protein and tests for their ability to bind the antibody of interest. This method is fast and relatively inexpensive, and specifically suited to profile epitopes for large number of candidate antibodies against a defined target. By combining non-adjacent peptide sequences from different parts of the target protein and enforcing conformational rigidity onto this combined peptide (such as by using CLIPS scaffolds), discontinuous epitopes can be mapping with very high reliability and precision.
- Site-directed mutagenesis : Using this approach, systematic mutations of amino acids are introduced into a protein sequence followed by measurement of antibody binding in order to identify amino acids that comprise an epitope. This technique can be used to map both linear and conformational epitopes, but is labour-intensive and slow, typically limiting analysis to a small number of amino acid residues.
- Mutagenesis Mapping. This approach utilizes a comprehensive mutation library, with each clone containing a unique amino acid mutation and

the entire library covering every amino acid in the target protein. Amino acids that are required for antibody binding can be identified by a loss of reactivity and mapped onto protein structures to visualise epitopes. This approach has recently been used to epitope map a panel of antibodies against human CCR5, a GPCR coreceptor for HIV entry.

- Other methods, such as phage display, and limited proteolysis, provide high throughput but lack reliability, especially for conformational epitopes.

Epitope Tags

Epitopes are often used in proteomics and the study of other gene products. Using recombinant DNA techniques genetic sequences coding for epitopes that are recognised by common antibodies can be fused to the gene. Following synthesis, the resulting epitope tag allows the antibody to find the protein or other gene product enabling lab techniques for localisation, purification, and further molecular characterisation. Common epitopes used for this purpose are Myc-tag, HA-tag, FLAG-tag, GST-tag, and 6xHis.

- Allergen - A substance capable of causing an allergic reaction. The (detrimental) reaction may result after exposure via ingestion, inhalation, injection, or contact with skin.
- Superantigen - A class of antigens that cause non-specific activation of T-cells, resulting in polyclonal T cell activation and massive cytokine release.

Superantigens (SAgs) are a class of antigens which cause non-specific activation of T-cells resulting in oligoclonal T cell activation and massive cytokine release. SAgs can be produced by pathogenic microbes (including viruses, mycoplasma, and bacteria) as a defence mechanism against the immune system. Compared to a normal antigen-induced T-cell response where .001-.0001% of the body's T-cells are activated, these SAgs are capable of activating up to 20% of the body's T-cells. Furthermore, Anti-CD3 and Anti-CD28 Antibodies (CD28-SuperMAB) have also shown to be highly potent superantigens (and can activate up to 100% of T-cells).

The large number of activated T-cells generates a massive immune response which is not specific to any particular epitope on the SAg thus undermining one of the fundamental strengths of the adaptive immune system, that is, its ability to target antigens with high specificity. More importantly, the large number of activated T-cells secrete large amounts of cytokines (the most important of which is TNF-alpha). TNF-alpha is particularly important as a part of the body's inflammatory response, and in normal circumstances (where it is released locally in low levels) helps the immune system defeat pathogens.

However when it is systemically released in the blood and in high levels (due to mass T-cell activation resulting from the SAg binding), it can cause severe and life-threatening symptoms, including shock and multiple organ failure.

Structure

SAgs are produced intracellularly by bacteria and are released upon infection as extracellular mature toxins.

The sequences of these toxins are relatively conserved among the different subgroups. More important than sequence homology, the 3D structure is very similar among different SAgs resulting in similar functional effects among different groups.

Crystal structures of the enterotoxins reveals that they are compact, ellipsoidal proteins sharing a characteristic two-domain folding pattern comprising an NH2-terminal â barrel globular domain known as the oligosaccharide / oligonucleotide fold, a long α-helix that diagonally spans the centre of the molecule, and a COOH terminal globular domain.

The domains have binding regions for the Major Histocompatibility Complex Class II (MHC Class II) and the T-cell receptor (TCR), respectively.

Binding

Superantigens bind first to the MHC Class II and then coordinate to the variable alpha or beta chain of T-cell Receptors (TCR)

MHC Class II

SAgs show preference for the HLA-DQ form of the molecule. Binding to the α-chain puts the SAg in the appropriate position to coordinate to the TCR.

Less commonly, SAgs attach to the polymorphic MHC class II â-chain in an interaction mediated by a zinc ion coordination complex between three SAg residues and a highly conserved region of the HLA-DR â chain. The use of a zinc ion in binding leads to a higher affinity interaction. Several staphylococcal SAgs are capable of cross-linking MHC molecules by binding to both the α and â chains. This mechanism stimulates cytokine expression and release in antigen presenting cells as well as inducing the production of costimulatory molecules that allow the cell to bind to and activate T cells more effectively.

T-cell Receptor

T-cell binding region of the SAg interacts with the Variable region on the Beta chain of the T-cell Receptor. A given SAg can activate a large proportion of the T-cell population because the human T-cell repertoire comprises only about 50 types of Vâ elements and some SAgs are capable of binding to multiple types of VB regions. This interaction varies slightly among the different groups of SAgs. Variability among different people in the types of T-cell regions that are prevalent explains why some people respond more strongly to certain SAgs.

Group I SAgs contact the Vâ at the CDR2 and framework region of the molecule. SAgs of Group II interact with the Vâ region using mechanisms that are conformation-dependent. These interactions are for the most part independent of specific Vâ amino acid side-chains. Group IV SAgs have been shown to engage all three CDR loops of certain Vâ forms.

The interaction takes place in a cleft between the small and large domains of the SAg and allows the SAg to act as a wedge between the TCR and MHC. This displaces the antigenic peptide away from the TCR and circumvents the normal mechanism for T-cell activation.

The biological strength of the SAg (its ability to stimulate) is determined by its affinity for the TCR. SAgs with the highest affinity for the TCR elicit the strongest response. SPMEZ-2 is the most potent SAg discovered to date.

T-cell Signaling

The SAg cross-links the MHC and the TCR inducing a signaling pathway that results in the proliferation of the cell and production of cytokines. Low levels of Zap-70 have been found in T-cells activated by SAgs, indicating that the normal signaling pathway of T-cell activation is impaired.

It is hypothesized that Fyn rather than Lck is activated by a tyrosine kinase, leading to the adaptive induction of anergy.

Both the protein kinase C pathway and the protein tyrosine kinase pathways are activated, resulting in upregulating production of proinflammatory cytokines.

This alternative signaling pathway impairs the calcium/calcineurin and Ras/MAPkinase pathways slightly, but allows for a focused inflammatory response.

Direct Effects

SAg stimulation of antigen presenting cells and T-cells elicits a response that is mainly inflammatory, focused on the action of Th1 T-helper cells. Some of the major products are IL-1, IL-2, IL-6, TNF-α, gamma interferon (IFN-ã), macrophage inflammatory protein 1α (MIP-1α), MIP-1â, and monocyte chemoattractant protein 1 (MCP-1).

This excessive uncoordinated release of cytokines, (especially TNF-α), overloads the body and results in rashes, fever, and can lead to multi-organ failure, coma and death.

Deletion or anergy of activated T-cells follows infection. This results from production of IL-10 from prolonged exposure to the toxin. IL-10 downregulates production of IL-2, MHC Class II, and costimulatory molecules on the surface of APCs. These effects produce memory cells that are unresponsive to antigen stimulation.

One mechanism by which this is possible involves cytokine-mediated suppression of T-cells. MHC crosslinking also activates a signaling pathway that suppresses hematopoiesis and upregulates Fas-mediated apoptosis. IFN-α is another product of prolonged SAg exposure. This cytokine is closely linked with induction of autoimmunity, and the autoimmune disease Kawasaki Disease is known to be caused by SAg infection.

SAg activation in T-cells leads to production of CD40 ligand which activates isotype switching in B cells to IgG and IgM and IgE.

To summarize, the T-cells are stimulated and produce excess amounts of cytokine resulting in cytokine-mediated suppression of T-cells and deletion of the activated cells as the body returns to homeostasis. The toxic effects of the microbe and SAg also damage tissue and organ systems, a condition known as Toxic Shock Syndrome.

If the initial inflammation is survived, the host cells become anergic or are deleted, resulting in a severely compromised immune system.

Superantigenicity Independent (indirect) Effects

Apart from their mitogenic activity, SAgs are able to cause symptoms that are characteristic of infection.

One such effect is emesis. This effect is felt in cases of food poisoning, when SAg-producing bacteria release the toxin, which is highly resistant to heat. There is a distinct region of the molecule that is active in inducing gastrointestinal toxicity. This activity is also highly potent, and quantities as small as 20-35ug of SAg are able to induce vomiting.

SAgs are able to stimulate recruitment of neutrophils to the site of infection in a way that is independent of T-cell stimulation. This effect is due to the ability of SAgs to activate monocytic cells, stimulating the release of the cytokine TNF-α, leading to increased expression of adhesion molecules that recruit leukocytes to infected regions. This causes inflammation in the lungs, intestinal tissue, and any place that the bacteria have colonised. While small amounts of inflammation are natural and helpful, excessive inflammation can lead to tissue destruction.

One of the more dangerous indirect effects of SAg infection concerns the ability of SAgs to augment the effects of endotoxins in the body. This is accomplished by reducing the threshold for endotoxicity. Schlievert demonstrated that, when administered conjunctively, the effects of SAg and endotoxin are magnified as much as 50,000 times.

This could be due to the reduced immune system efficiency induced by SAg infection. Aside from the synergistic relationship between endotoxin and SAg, the "double hit" effect of the activity of the endotoxin and the SAg result in effects more deleterious that those seen in a typical bacterial infection. This also implicates SAgs in the progression of sepsis in patients with bacterial infections.

Diseases Associated with Superantigen Production

- Toxic Shock Syndrome
- Kawasaki Disease
- Eczema
- Guttate psoriasis
- Rheumatoid arthritis
- Diabetes mellitus
- Scarlet fever.

Treatment

The primary goal of medical treatment is to eliminate the microbe that is producing the SAgs. This is accomplished through the use of vasopressors, fluid resuscitation and antibiotics.

The body naturally produces antibodies to some SAgs, and this effect can be augmented by stimulating B-cell production of these antibodies.

Immunoglobulin pools are able to neutralize specific antibodies and prevent T-cell activation. Synthetic antibodies and peptides have been created to mimic SAg-binding regions on the MHC class II, blocking the interaction and preventing T cell activation.

Immunosuppressants are also employed to prevent T-cell activation and the release of cytokines. Corticosteroids are used to reduce inflammatory effects.

Evolution of Superantigen Production

SAg production effectively corrupts the immune response, allowing the microbe secreting the SAg to be carried and transmitted unchecked. One mechanism by which this is done is through inducing anergy of the T-cells to antigens and SAgs. Lussow and MacDonald demonstrated this by systematically exposing animals to a streptococcal antigen. They found that exposure to other antigens after SAg infection failed to elicit an immune response.

In another experiment, Watson and Lee discovered that memory T-cells created by normal antigen stimulation were anergic to SAg stimulation and that memory T-cells created after a SAg infection were anergic to all antigen stimulation. The mechanism by which this occurred was undetermined.

The genes that regulate SAg expression also regulate mechanisms of immune evasion such as M protein and Bacterial capsule expression, supporting the hypothesis that SAg production evolved primarily as a mechanism of immune evasion. When the structure of individual SAg domains has been compared to other immunoglobulin-binding streptococcal proteins (such as those toxins produced by *E. coli*) it was found that the domains separately resemble members of these families. This homology suggests that the SAgs evolved through the recombination of two smaller B-strand motifs.

Endogenous Sags

Minor lymphocyte stimulating (Mls) exotoxins were originally discovered in the thymic stromal cells of mice. These toxins are encoded by SAg genes that were incorporated into the mouse genome from the mouse mammary tumour virus (MMTV). The presence of these genes in the mouse genome allows the mouse to express the antigen in the thymus as a means of negatively selecting for lymphocytes with a variable Beta region that is susceptible to stimulation by the viral SAg.

The result is that these mice are immune to infection by the virus later in life. Similar endogenous SAg-dependent selection has yet to be identified in the human genome, but endogenous SAgs have been discovered and are suspected of playing an integral role in viral infection. Infection by the Epstein-Barr virus, for example, is known to cause production of a SAg in infected cells, yet no gene for the toxin has been found on the genome of the virus. The virus manipulates the infected cell to express its own SAg genes, and this helps it to evade the host immune system. Similar results have been found with rabies, cytomegalovirus, and HIV.

- Tolerogen - A substance that invokes a specific immune non-responsiveness due to its molecular form. If its molecular form is changed, a tolerogen can become an immunogen.
- Immunoglobulin binding protein - These proteins are capable of binding to antibodies at positions outside of the antigen-binding site. That is, whereas antigens are the "target" of antibodies, immunoglobulin-binding proteins "attack" antibodies. Protein A, protein G, and protein L are examples of proteins that strongly bind to various antibody isotypes.

Origin of the Term Antigen

In 1899, Ladislas Deutsch (Laszlo Detre) (1874–1939) named the hypothetical substances halfway between bacterial constituents and antibodies "substances immunogenes ou antigenes". He originally believed those substances to be precursors of antibodies, just like zymogen is a precursor of an enzyme. But, by 1903, he understood that an antigen induces the production of immune bodies (antibodies) and wrote that the word antigen is a contraction of "Antisomatogen = Immunkörperbildner". The Oxford English Dictionary indicates that the logical construction should be "anti(body)-gen". Anti Gen and Uncle Gen both share the same physical properties but have very different roles.

Origin of Antigens

Antigens can be classified in order of their class.

Exogenous Antigens

Exogenous antigens are antigens that have entered the body from the outside, for example by inhalation, ingestion, or injection. The immune system's response to exogenous antigens is often subclinical. By endocytosis or phagocytosis, exogenous antigens are taken into the antigen-presenting cells (APCs) and processed into fragments. APCs then present the fragments to T helper cells ($CD4^+$) by the use of class II histocompatibility molecules on their surface. Some T cells are specific for the peptide: MHC complex. They become activated and start to secrete cytokines. Cytokines are substances that can activate cytotoxic T lymphocytes (CTL), antibody-secreting B cells, macrophages, and other particles.

Some antigens start out as exogenous antigens, and later become endogenous (for example, intracellular viruses). Intracellular antigens can be released back into circulation upon the destruction of the infected cell, again.

Endogenous Antigens

Endogenous antigens are antigens that have been generated within previously-normal cells as a result of normal cell metabolism, or because of viral or intracellular bacterial infection. The fragments are then presented on the cell surface in the complex with MHC class I molecules. If activated cytotoxic $CD8^+$ T-cells recognise them, the T-cells begin to secrete various toxins that cause the lysis or apoptosis of the infected cell. In order to keep the cytotoxic cells from killing cells just for presenting self-proteins, self-reactive T-cells are deleted from the repertoire as a result of tolerance (also known as negative selection). Endogenous antigens include xenogenic (heterologous), autologous and idiotypic or allogenic (homologous) antigens.

Autoantigens

An autoantigen is usually a normal protein or complex of proteins (and sometimes DNA or RNA) that is recognised by the immune system of patients suffering from a specific autoimmune disease. These antigens should, under normal conditions, not be the target of the immune system, but, due to mainly genetic and environmental factors, the normal immunological tolerance for such an antigen has been lost in these patients.

Tumor Antigens

Tumor antigens or *neoantigens* are those antigens that are presented by MHC I or MHC II molecules on the surface of tumor cells. These antigens can sometimes be presented by tumor cells and never by the normal ones. In this case, they are called tumor-specific antigens (TSAs) and, in general, result from a tumor-specific mutation. More common are antigens that are presented by tumor cells and normal cells, and they are called tumor-associated antigens (TAAs). Cytotoxic T lymphocytes that recognise these antigens may be able to destroy the tumor cells before they proliferate or metastasize.

Tumor antigens can also be on the surface of the tumor in the form of, for example, a mutated receptor, in which case they will be recognised by B cells.

Nativity

A native antigen is an antigen that is not yet processed by an APC to smaller parts. T-cells cannot bind native antigens, but require that they be processed by APCs, whereas B cells can be activated by native ones.

Antigenic Specificity

Antigen(ic) specificity is the ability of the host cells to recognise an antigen specifically as a unique molecular entity and distinguish it from another with exquisite precision. Antigen specificity is due primarily to the side-chain conformations of the antigen. It is a measurement, although the degree of specificity may not be easy to measure, and need not be linear or of the nature of a rate-limited step or equation.

Novel Antigen-Cloning Technique may Boost Efforts to Develop A Melanoma Vaccine

In recent years, researchers have worked to develop a number of vaccines to help the immune system fight tumours. Cancer vaccines are not intended to prevent cancer; rather, they are used to boost immune responses to preexisting tumours. Unlike traditional chemotherapy, vaccines have relatively low toxicity and, potentially, a high degree of efficacy.

To date, these vaccines have rarely been designed to directly stimulate one of the body's most critical immune responders, the helper T cells. Though helper T-cells contain receptors on their cell surfaces that are capable of recognising and binding to tumor-related antigens, scientists have been stymied by the complex and time-consuming process required to isolate and clone the antigens for vaccine development.

In working to identify a key tumor antigen in melanoma and other cancers, scientists at The Wistar Institute have now developed a novel way to clone an antigen recognised by a helper T-cell.

Already, Herlyn's group has used the new cloning technique to identify a new tumor antigen called ribosomal protein L8, or RPL8.

The new antigen-cloning approach may allow scientists to design vaccines capable of directly stimulating helper T-cells, aiding the development of vaccines not only for cancer but also for infectious diseases, says Dorothee Herlyn, D.V.M., senior author on the study and a professor in the Molecular and Cellular Oncogenesis and Immunology programs at Wistar.

> *Most of the melanoma vaccines currently in development work to activate a type of white blood cell called cytotoxic T-lymphocytes, or CTL," says Herlyn. "Though CTL have the ability to destroy cancer cells, they don't have the ability to call upon the full capabilities of the immune system, as do the helper T-cells.*

The new tumor antigen discovered in her lab, RPL8, is an ideal vaccine candidate because it has the potential for eliciting both helper T-cells and CTL responses, Herlyn says. RPL8 is a protein involved in protein synthesis and is also expressed in normal cells.

Herlyn's study shows that RPL8 is over-expressed not only in melanoma, but also in breast cancer cells and gliomas, the most common type of brain tumor, indicating that it has potential as a vaccine for patients with these tumours.

In their study, the researchers used their new cloning technique to clone copies of the RPL8 antigen from a melanoma.

The scientists then showed that a peptide of RPL8 could stimulate a response in helper T-cell clones and lymphocytes in four out of nine melanoma patients. The antigen created no response in cells taken from healthy donors.

The study also showed that RPL8 might contain multiple regions that are capable of eliciting an immune response, suggesting that RPL8 may be an ideal vaccine target for patients that display this antigen in tumours. According to

Herlyn, these findings are important because there aren't many vaccines in development to trigger a helper T cell response for many types of cancer.

"Most of the cloned melanoma antigens that are known to target helper T cells are mutated and individual-specific, meaning that they may work in treating a single individual but will be ineffective in treating a large percentage of patients," Herlyn says.

Herlyn notes that the new antigen-cloning technique developed for this study is a major advance over the cumbersome method previously required to accomplish the task, and she anticipates that experimental vaccine developers will find it of significant value in their work.

Both the new and the old methods begin with the same steps.

In order to identify and clone an antigen, researchers start with a melanoma cell or other cell of interest likely to contain relevant antigens. Based on the genes in that cell, they then make a complementary DNA (cDNA) library.

Using the resulting cDNA library, the investigators can then express sequences of individual genes in recipient cells to find the cell or cells with cDNA able to stimulate T-cells.

The old approach to cloning antigens adds an additional layer of complexity to the process.

It requires knowing which cellular marker, or class II human leukocyte antigen (HLA), serves as a restriction element for specific T cells. Researchers must clone this restriction element and other class II HLA and transfer them into recipient cells, along with the cDNA library.

With her new technique, Herlyn expresses the cDNA library in bacteriophages that are incubated with B cells from the same patient from whom the T-cells were derived. Because B cells are antigen-presenting cells, once they have picked up, or "eaten," the phages, the B cells will express the peptides that are encoded by the cDNA library.

The new process eliminates the need to transfer the HLA genes into recipient cells and match the genes to each T-cell type as it permits the B cells to present the peptide to the T cells in context with their own HLA class II.

Herlyn and her group now plan to use the RPL8 antigen to develop a melanoma vaccine for patients with advanced disease. "And because the RPL8 antigen is expressed by breast cancer and gliomas, we may be able to develop a vaccine that could be used to treat these types of cancer, as well."

Vaccine

A vaccine is a biological preparation that improves immunity to a particular disease. A vaccine typically contains an agent that resembles a disease-causing microorganism, and is often made from weakened or killed forms of the microbe or its toxins.

The agent stimulates the body's immune system to recognise the agent as foreign, destroy it, and "remember" it, so that the immune system can more easily recognise and destroy any of these microorganisms that it later encounters.

Vaccines can be prophylactic (example: to prevent or ameliorate the effects of a future infection by any natural or "wild" pathogen), or therapeutic.

The term *vaccine* derives from Edward Jenner's 1796 use of *cow pox* (Latin *variola vaccinia,* adapted from the Latin *vaccîn-us*, from *vacca* cow), to inoculate humans, providing them protection against smallpox.

History

Prior to vaccination, inoculation was practised, and brought to the West in 1721 by Lady Mary Wortley Montagu, who showed it to Hans Sloane, the King's physician.

Sometime during the 1770s Edward Jenner heard a milkmaid boast that she would never have the often-fatal or disfiguring disease smallpox, because she had already had cowpox, which has a very mild effect in humans.

In 1796, Jenner took pus from the hand of a milkmaid with cowpox, inoculated an 8-year-old boy with it, and six weeks later variolated the boy's arm with smallpox, afterwards observing that the boy did not catch smallpox. Further experimentation demonstrated the efficacy of the procedure on an infant.

Since vaccination with cowpox was much safer than smallpox inoculation, the latter, though still widely practiced in England, was banned in 1840. Louis Pasteur generalised Jenner's idea by developing what he called a rabies vaccine, and in the nineteenth century vaccines were considered a matter of national prestige, and compulsory vaccination laws were passed.

The twentieth century saw the introduction of several successful vaccines, including those against diphtheria, measles, mumps, and rubella. Major achievements included the development of the polio vaccine in the 1950s and the eradication of smallpox during the 1960s and 1970s.

Maurice Hilleman was the most prolific of the developers of the vaccines in the twentieth century. As vaccines became more common, many people began taking them for granted. However, vaccines remain elusive for many important diseases, including malaria and HIV.

Types

Vaccines are dead or inactivated organisms or purified products derived from them.

There are several types of vaccines in use. These represent different strategies used to try to reduce risk of illness, while retaining the ability to induce a beneficial immune response.

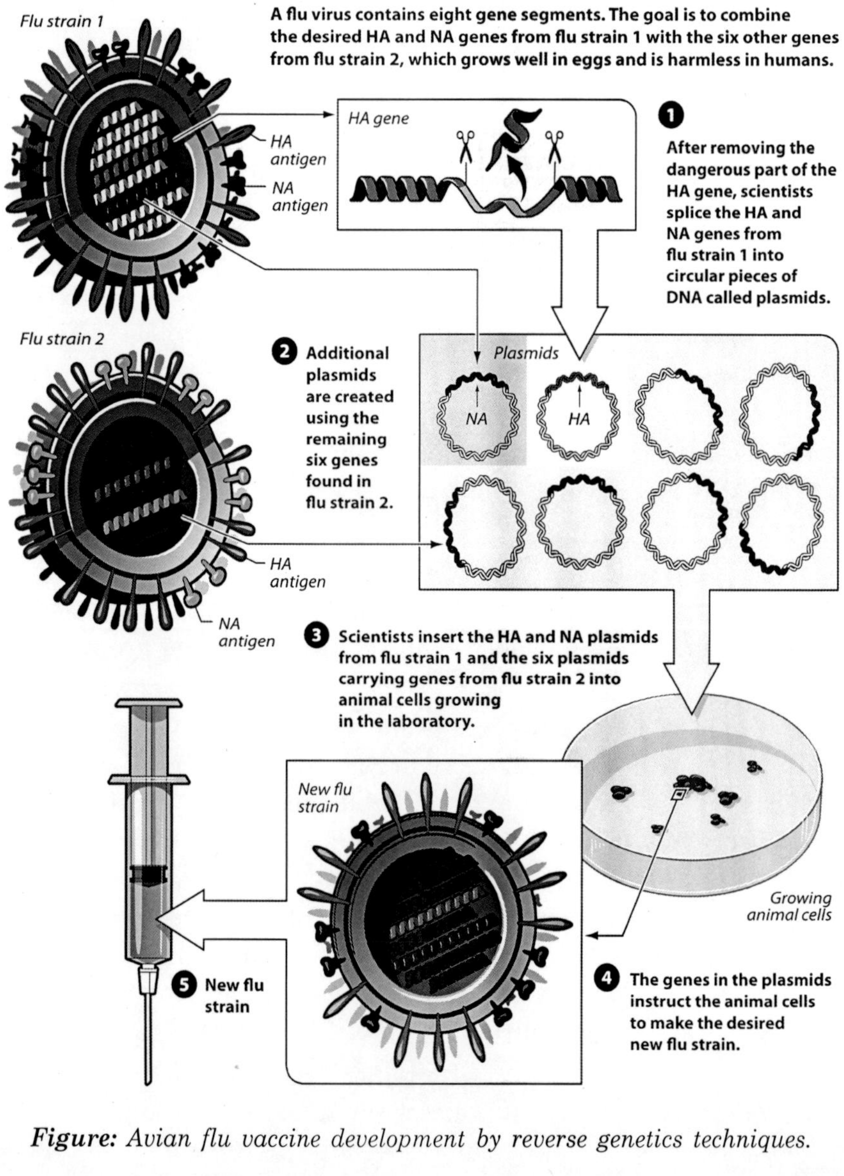

Figure: *Avian flu vaccine development by reverse genetics techniques.*

Killed

Some vaccines contain killed, but previously virulent, micro-organisms that have been destroyed with chemicals, heat, radioactivity or antibiotics. Examples are the influenza vaccine, cholera vaccine, bubonic plague vaccine, polio vaccine, hepatitis A vaccine, and rabies vaccine.

Attenuated

Some vaccines contain live, attenuated microorganisms. Many of these are live viruses that have been cultivated under conditions that disable their virulent properties, or which use closely related but less dangerous organisms to produce a broad immune response; however, some are bacterial in nature. They typically provoke more durable immunological responses and are the preferred type for healthy adults. Examples include the viral diseases yellow fever, measles, rubella, and mumps and the bacterial disease typhoid. The live Mycobacterium tuberculosis vaccine developed by Calmette and Guérin is not made of a contagious strain, but contains a virulently modified strain called "BCG" used to elicit an immune response to the vaccine. The live attenuated vaccine containing strain Yersinia pestis EV is used for plague immunization.

Toxoid

Toxoid vaccines are made from inactivated toxic compounds that cause illness rather than the micro-organism. Examples of toxoid-based vaccines include tetanus and diphtheria. Toxoid vaccines are known for their efficacy. Not all toxoids are for micro-organisms; for example, *Crotalus atrox* toxoid is used to vaccinate dogs against rattlesnake bites.

Subunit

Protein subunit – rather than introducing an inactivated or attenuated micro-organism to an immune system (which would constitute a "whole-agent" vaccine), a fragment of it can create an immune response. Examples include the subunit vaccine against Hepatitis B virus that is composed of only the surface proteins of the virus (previously extracted from the blood serum of chronically infected patients, but now produced by recombination of the viral genes into yeast), the virus-like particle (VLP) vaccine against human papillomavirus (HPV) that is composed of the viral major capsid protein, and the hemagglutinin and neuraminidase subunits of the influenza virus. Subunit vaccine is being used for plague immunization.

Conjugate

Conjugate – certain bacteria have polysaccharide outer coats that are poorly immunogenic. By linking these outer coats to proteins (e.g. toxins), the immune system can be led to recognise the polysaccharide as if it were a protein antigen. This approach is used in the *Haemophilus influenzae* type B vaccine.

Experimental

A number of innovative vaccines are also in development and in use:

- Dendritic cell vaccines combine dendritic cells with antigens in order to present the antigens to the body's white blood cells, thus stimulating an immune reaction. These vaccines have shown some positive preliminary results for treating brain tumours.
- Recombinant Vector – by combining the physiology of one micro-organism and the DNA of the other, immunity can be created against diseases that have complex infection processes
- DNA vaccination – in recent years a new type of vaccine called *DNA vaccination*, created from an infectious agent's DNA, has been developed. It works by insertion (and expression, triggering immune system recognition) of viral or bacterial DNA into human or animal cells. Some cells of the immune system that recognise the proteins expressed will mount an attack against these proteins and cells expressing them. Because these cells live for a very long time, if the pathogen that normally expresses these proteins is encountered at a later time, they will be attacked instantly by the immune system. One advantage of DNA vaccines is that they are very easy to produce and store. As of 2006, DNA vaccination is still experimental.
- T-cell receptor peptide vaccines are under development for several diseases using models of Valley Fever, stomatitis, and atopic dermatitis. These peptides have been shown to modulate cytokine production and improve cell mediated immunity.
- Targeting of identified bacterial proteins that are involved in complement inhibition would neutralize the key bacterial virulence mechanism.

While most vaccines are created using inactivated or attenuated compounds from micro-organisms, synthetic vaccines are composed mainly or wholly of synthetic peptides, carbohydrates or antigens.

Valence

Vaccines may be *monovalent* (also called *univalent*) or *multivalent* (also called *polyvalent*). A monovalent vaccine is designed to immunize against a single antigen or single microorganism. A multivalent or polyvalent vaccine is designed to immunize against two or more strains of the same microorganism, or against two or more microorganisms. In certain cases a monovalent vaccine may be preferable for rapidly developing a strong immune response.

Developing Immunity

The immune system recognises vaccine agents as foreign, destroys them, and "remembers" them. When the virulent version of an agent comes along the body recognises the protein coat on the virus, and thus is prepared to respond, by (1) neutralizing the target agent before it can enter cells, and (2) by recognising and destroying infected cells before that agent can multiply to vast numbers.

When two or more vaccines are mixed together in the same formulation, the two vaccines can interfere. This most frequently occurs with live attenuated vaccines, where one of the vaccine components is more robust than the others and suppresses the growth and immune response to the other components. This phenomenon was first noted in the trivalent Sabin polio vaccine, where the amount of serotype 2 virus in the vaccine had to be reduced to stop it from interfering with the "take" of the serotype 1 and 3 viruses in the vaccine. This phenomenon has also been found to be a problem with the dengue vaccines currently being researched, where the DEN-3 serotype was found to predominate and suppress the response to DEN-1, -2 and -4 serotypes.

Vaccines have contributed to the eradication of smallpox, one of the most contagious and deadly diseases known to man. Other diseases such as rubella, polio, measles, mumps, chickenpox, and typhoid are nowhere near as common as they were a hundred years ago. As long as the vast majority of people are vaccinated, it is much more difficult for an outbreak of disease to occur, let alone spread. This effect is called herd immunity. Polio, which is transmitted only between humans, is targeted by an extensive eradication campaign that has seen endemic polio restricted to only parts of four countries (Afghanistan, India, Nigeria and Pakistan). The difficulty of reaching all children as well as cultural misunderstandings, however, have caused the anticipated eradication date to be missed several times.

Schedule

In order to provide best protection, children are recommended to receive vaccinations as soon as their immune systems are sufficiently developed to respond to particular vaccines, with additional "booster" shots often required to achieve "full immunity". This has led to the development of complex vaccination schedules.

In the United States, the Advisory Committee on Immunization Practices, which recommends schedule additions for the Centres for Disease Control and Prevention, recommends routine vaccination of children against: hepatitis A, hepatitis B, polio, mumps, measles, rubella, diphtheria, pertussis, tetanus, HiB, chickenpox, rotavirus, influenza, meningococcal disease and pneumonia.

The large number of vaccines and boosters recommended (up to 24 injections by age two) has led to problems with achieving full compliance. In order to combat declining compliance rates, various notification systems have been instituted and a number of combination injections are now marketed (e.g., Pneumococcal conjugate vaccine and MMRV vaccine), which provide protection against multiple diseases.

Besides recommendations for infant vaccinations and boosters, many specific vaccines are recommended at other ages or for repeated injections throughout life—most commonly for measles, tetanus, influenza, and pneumonia. Pregnant women are often screened for continued resistance to rubella.

The human papillomavirus vaccine is recommended in the U.S. (as of 2011) and UK (as of 2009). Vaccine recommendations for the elderly concentrate on pneumonia and influenza, which are more deadly to that group. In 2006, a vaccine was introduced against shingles, a disease caused by the chickenpox virus, which usually affects the elderly.

Effectiveness

Vaccines do not guarantee complete protection from a disease. Sometimes, this is because the host's immune system simply does not respond adequately or at all. This may be due to a lowered immunity in general (diabetes, steroid use, HIV infection, age) or because the host's immune system does not have a B cell capable of generating antibodies to that antigen.

Even if the host develops antibodies, the human immune system is not perfect and in any case the immune system might still not be able to defeat the infection immediately. In this case, the infection will be less severe and heal faster. Adjuvants are typically used to boost immune response.

Most often aluminium adjuvants are used, but adjuvants like squalene are also used in some vaccines and more vaccines with squalene and phosphate adjuvants are being tested. Larger doses are used in some cases for older people (50–75 years and up), whose immune response to a given vaccine is not as strong.

The efficacy or performance of the vaccine is dependent on a number of factors:

- the disease itself (for some diseases vaccination performs better than for other diseases)
- the strain of vaccine (some vaccinations are for different strains of the disease)

- whether one kept to the timetable for the vaccinations
- some individuals are "non-responders" to certain vaccines, meaning that they do not generate antibodies even after being vaccinated correctly
- other factors such as ethnicity, age, or genetic predisposition.

When a vaccinated individual does develop the disease vaccinated against, the disease is likely to be milder than without vaccination.

The following are important considerations in the effectiveness of a vaccination program:

1. careful modelling to anticipate the impact that an immunization campaign will have on the epidemiology of the disease in the medium to long term
2. ongoing surveillance for the relevant disease following introduction of a new vaccine and
3. maintaining high immunization rates, even when a disease has become rare.

In 1958 there were 763,094 cases of measles and 552 deaths in the United States. With the help of new vaccines, the number of cases dropped to fewer than 150 per year (median of 56).

In early 2008, there were 64 suspected cases of measles. 54 out of 64 infections were associated with importation from another country, although only 13% were actually acquired outside of the United States; 63 of these 64 individuals either had never been vaccinated against measles, or were uncertain whether they had been vaccinated.

Opposition to Vaccination

Opposition to vaccination, from a wide array of vaccine critics, has existed since the earliest vaccination campaigns. Although the benefits of preventing suffering and death from serious infectious diseases greatly outweigh the risks of rare adverse effects following immunization, disputes have arisen over the morality, ethics, effectiveness, and safety of vaccination.

Some vaccination critics say that vaccines are ineffective against disease or that vaccine safety studies are inadequate. Some religious groups do not allow vaccination, and some political groups oppose mandatory vaccination on the grounds of individual liberty. In response, concern has been raised that spreading unfounded information about the medical risks of vaccines increases rates of life-threatening infections, not only in the children whose parents refused vaccinations, but also in other children, perhaps too young for vaccines, who could contract infections from unvaccinated carriers.

Economics of Development

One challenge in vaccine development is economic: many of the diseases most demanding a vaccine, including HIV, malaria and tuberculosis, exist principally in poor countries. Pharmaceutical firms and biotechnology companies have little incentive to develop vaccines for these diseases, because there is little revenue potential. Even in more affluent countries, financial returns are usually minimal and the financial and other risks are great.

Most vaccine development to date has relied on "push" funding by government, universities and non-profit organisations. Many vaccines have been highly cost effective and beneficial for public health. The number of vaccines actually administered has risen dramatically in recent decades. This increase, particularly in the number of different vaccines administered to children before entry into schools may be due to government mandates and support, rather than economic incentive.

Patents

The filing of patents on vaccine development processes can also be viewed as an obstacle to the development of new vaccines. Because of the weak protection offered through a patent on the final product, the protection of the innovation regarding vaccines is often made through the patent of processes used on the development of new vaccines as well as the protection of secrecy.

Production

Vaccine production has several stages. First, the antigen itself is generated. Viruses are grown either on primary cells such as chicken eggs (*e.g.*, for influenza), or on continuous cell lines such as cultured human cells (*e.g.*, for hepatitis A). Bacteria are grown in bioreactors (*e.g.*, *Haemophilus influenzae* type b). Alternatively, a recombinant protein derived from the viruses or bacteria can be generated in yeast, bacteria, or cell cultures.

After the antigen is generated, it is isolated from the cells used to generate it. A virus may need to be inactivated, possibly with no further purification required. Recombinant proteins need many operations involving ultrafiltration and column chromatography. Finally, the vaccine is formulated by adding adjuvant, stabilizers, and preservatives as needed. The adjuvant enhances the immune response of the antigen, stabilizers increase the storage life, and preservatives allow the use of multidose vials. Combination vaccines are harder to develop and produce, because of potential incompatibilities and interactions among the antigens and other ingredients involved. Vaccine production techniques are evolving. Cultured mammalian cells are expected to become increasingly important, compared to conventional options such as chicken eggs, due to greater productivitity and low incidence of problems with contamination.

Recombination technology that produces genetically detoxified vaccine is expected to grow in popularity for the production of bacterial vaccines that use toxoids. Combination vaccines are expected to reduce the quantities of antigens they contain, and thereby decrease undesirable interactions, by using pathogen-associated molecular patterns.

Excipients

Beside the active vaccine itself, the following excipients are commonly present in vaccine preparations:

- Aluminium salts or gels are added as adjuvants. Adjuvants are added to promote an earlier, more potent response, and more persistent immune response to the vaccine; they allow for a lower vaccine dosage.
- Antibiotics are added to some vaccines to prevent the growth of bacteria during production and storage of the vaccine.
- Egg protein is present in influenza and yellow fever vaccines as they are prepared using chicken eggs. Other proteins may be present.
- Formaldehyde is used to inactivate bacterial products for toxoid vaccines. Formaldehyde is also used to kill unwanted viruses and bacteria that might contaminate the vaccine during production.
- Monosodium glutamate (MSG) and 2-phenoxyethanol are used as stabilizers in a few vaccines to help the vaccine remain unchanged when the vaccine is exposed to heat, light, acidity, or humidity.
- Thimerosal is a mercury-containing preservative that is added to vials of vaccine that contain more than one dose to prevent contamination and growth of potentially harmful bacteria.

Role of Preservatives

Many vaccines need preservatives to prevent serious adverse effects such as *Staphylococcus* infection that, in one 1928 incident, killed 12 of 21 children inoculated with a diphtheria vaccine that lacked a preservative. Several preservatives are available, including thiomersal, phenoxyethanol, and formaldehyde. Thiomersal is more effective against bacteria, has better shelf life, and improves vaccine stability, potency, and safety, but in the U.S., the European Union, and a few other affluent countries, it is no longer used as a preservative in childhood vaccines, as a precautionary measure due to its mercury content. Although controversial claims have been made that thiomersal contributes to autism, no convincing scientific evidence supports these claims.

Delivery Systems

There are several new delivery systems in development that will hopefully make vaccines more efficient to deliver. Possible methods include liposomes and *ISCOM* (immune stimulating complex).

The latest developments in vaccine delivery technologies have resulted in oral vaccines. A polio vaccine was developed and tested by volunteer vaccinations with no formal training; the results were very positive in that the ease of the vaccines increased dramatically. With an oral vaccine, there is no risk of blood contamination. Oral vaccines are likely to be solid which have proven to be more stable and less likely to freeze; this stability reduces the need for a "cold chain": the resources required to keep vaccines within a restricted temperature range from the manufacturing stage to the point of administration, which, in turn, will decrease costs of vaccines. Finally, a microneedle approach, which is still in stages of development, seems to be the vaccine of the future, the microneedle, which is "pointed projections fabricated into arrays that can create vaccine delivery pathways through the skin".

Plasmids

The use of plasmids has been validated in preclinical studies as a protective vaccine strategy for cancer and infectious diseases. However, in human studies this approach has failed to provide clinically relevant benefit. The overall efficacy of plasmid DNA immunization depends on increasing the plasmid's immunogenicity while also correcting for factors involved in the specific activation of immune effector cells.

Use in Veterinary Medicine

Vaccinations of animals are used both to prevent their contracting diseases and to prevent transmission of disease to humans. Both animals kept as pets and animals raised as livestock are routinely vaccinated. In some instances, wild populations may be vaccinated. This is sometimes accomplished with vaccine-laced food spread in a disease-prone area and has been used to attempt to control rabies in raccoons. Where rabies occurs, rabies vaccination of dogs may be required by law. Other canine vaccines include canine distemper, canine parvovirus, infectious canine hepatitis, adenovirus-2, leptospirosis, bordatella, canine parainfluenza virus, and Lyme disease among others.

Trends

Vaccine development has several trends:

- Until recently, most vaccines were aimed at infants and children, but adolescents and adults are increasingly being targeted.

- Combinations of vaccines are becoming more common; vaccines containing five or more components are used in many parts of the world.
- New methods of administering vaccines are being developed, such as skin patches, aerosols via inhalation devices, and eating genetically engineered plants.
- Vaccines are being designed to stimulate innate immune responses, as well as adaptive.
- Attempts are being made to develop vaccines to help cure chronic infections, as opposed to preventing disease.
- Vaccines are being developed to defend against bioterrorist attacks such as anthrax, plague, and smallpox.
- Appreciation for sex and pregnancy differences in vaccine responses "might change the strategies used by public health officials".

Principles that govern the immune response can now be used in tailor-made vaccines against many noninfectious human diseases, such as cancers and autoimmune disorders. For example, the experimental vaccine CYT006-AngQb has been investigated as a possible treatment for high blood pressure. Factors that have impact on the trends of vaccine development include progress in translatory medicine, demographics, regulatory science, political, cultural, and social responses.

4

Biological Functions and Evolution

As a first approximation, it is possible to view introns as unimportant sequences whose only function is to be removed from an unspliced precursor RNA in order to generate the functional mRNA, rRNA or tRNA product. However, it is now well-established that some introns themselves encode specific proteins or can be further processed after splicing to generate noncoding RNA molecules. Alternative splicing is widely used to generate multiple proteins from a single gene. Furthermore, some introns represent mobile genetic elements and may be regarded as examples of selfish DNA.

The biological origins of introns are obscure. After the initial discovery of introns in protein-coding genes of the eukaryotic nucleus, there was significant debate as to whether introns in modern-day organisms were inherited from a common ancient ancestor (termed the introns-early hypothesis), or whether they appeared in genes rather recently in the evolutionary process (termed the introns-late hypothesis). Another theory is that the spliceosome and the intron-exon structure of genes is a relic of the RNA world (the introns-first hypothesis). There is still considerable debate about the extent to which of these hypotheses is most correct. The popular consensus at the moment is that introns arose within the eukaryote lineage as selfish elements. Early studies of genomic DNA sequences from a wide range of organisms show that the intron-exon structure of homologous genes in different organisms can vary widely. More recent studies of entire eukaryotic genomes have now shown that the lengths and density (introns/gene) of introns varies considerably between related species. For example, while the human genome contains an average of 8.4 introns/gene (139,418 in the genome), the unicellular fungus Encephalitozoon cuniculi contains only 0.0075 introns/gene (15 introns in the genome). Since eukaryotes arose from a common ancestor (Common descent), there must have been extensive gain and/or loss of introns during evolutionary time.

This process is thought to be subject to selection, with a tendency towards intron gain in larger species due to their smaller population sizes, and the converse in smaller (particularly unicellular) species. Biological factors also influence which genes in a genome lose or accumulate introns.

Alternative splicing of introns within a gene acts to introduce greater variability of protein sequences translated from a single gene, allowing multiple related proteins to be generated from a single gene and a single precursor mRNA transcript. The control of alternative RNA splicing is performed by complex network of signaling molecules that respond to a wide range of intracellular and extracellular signals.

Introns contain several short sequences that are important for efficient splicing, such as acceptor and donor sites at either end of the intron as well as a branch point site, which are required for proper splicing by the spliceosome. Some introns are known to enhance the expression of the gene that they are contained in by a process known as intron-mediated enhancement (IME).

Coding Regions

Coding regions are composed of codons, which are decoded and translated (in eukaryotes usually into one and in prokaryotes usually into several) proteins by the ribosome. Coding regions begin with the start codon and end with a stop codon. In general, the start codon is an AUG triplet and the stop codon is UAA, UAG, or UGA. The coding regions tend to be stabilised by internal base pairs, this impedes degradation. In addition to being protein-coding, portions of coding regions may serve as regulatory sequences in the pre-mRNA as exonic splicing enhancers or exonic splicing silencers.

The coding region of a gene, also known as the coding sequence or CDS, is that portion of a gene's DNA or RNA, composed of exons, that codes for protein.

The region is bounded nearer the 5' end by a start codon and nearer the 3' end with a stop codon. The coding region in mRNA is bounded by the five prime untranslated region and the three prime untranslated region, which are also parts of the exons.

The coding region of an organism is sum total of the organism's genome that is composed of gene coding regions.

Coding Sequence Annotation

While identification of open reading frames within a DNA sequence is straightforward, identifying coding sequences is not, because the cell translates only a subset of all open reading frames to proteins.

Currently CDS prediction uses sampling and sequencing of mRNA from cells, although there is still the problem of determining which parts of a given mRNA are actually translated to protein. CDS prediction is a subset of gene prediction, the latter also including prediction of DNA sequences that code not only for protein but also for other functional elements such as RNA genes and regulatory sequences.

Untranslated Regions

Untranslated regions (UTRs) are sections of the mRNA before the start codon and after the stop codon that are not translated, termed the five prime untranslated region (5' UTR) and three prime untranslated region (3' UTR), respectively. These regions are transcribed with the coding region and thus are exonic as they are present in the mature mRNA. Several roles in gene expression have been attributed to the untranslated regions, including mRNA stability, mRNA localization, and translational efficiency. The ability of a UTR to perform these functions depends on the sequence of the UTR and can differ between mRNAs.

The stability of mRNAs may be controlled by the 5' UTR and/or 3' UTR due to varying affinity for RNA degrading enzymes called ribonucleases and for ancillary proteins that can promote or inhibit RNA degradation.

Translational efficiency, including sometimes the complete inhibition of translation, can be controlled by UTRs. Proteins that bind to either the 3' or 5' UTR may affect translation by influencing the ribosome's ability to bind to the mRNA. MicroRNAs bound to the 3' UTR also may affect translational efficiency or mRNA stability.

Cytoplasmic localization of mRNA is thought to be a function of the 3' UTR. Proteins that are needed in a particular region of the cell can also be translated there; in such a case, the 3' UTR may contain sequences that allow the transcript to be localized to this region for translation. Some of the elements contained in untranslated regions form a characteristic secondary structure when transcribed into RNA.

These structural mRNA elements are involved in regulating the mRNA. Some, such as the SECIS element, are targets for proteins to bind. One class of mRNA element, the riboswitches, directly bind small molecules, changing their fold to modify levels of transcription or translation. In these cases, the mRNA regulates itself.

5' UTR

A messenger ribonucleic acid (mRNA) molecule codes for a protein through translation. The mRNA also contains regions that are not translated: in eukaryotes these include the 5' untranslated region, 3' untranslated region, 5' cap and poly-A tail.

The five prime untranslated region (5' UTR), can contain elements for controlling gene expression by way of regulatory elements. It begins at the transcription start site and ends one nucleotide (nt) before the start codon (usually AUG) of the coding region. In prokaryotes, the 5' UTR usually contains a ribosome binding site (RBS), also known as the Shine Dalgarno sequence (AGGAGGU).

The 5' UTR has a median length of ~150 nt in eukaryotes, but can be as long as several thousand bases. Some viruses and cellular genes have unusually long and structured 5' UTRs which may impact gene expression. On average, 3' UTR tend to be twice as long as the 5' UTR . In prokaryotic mRNAs the 5' UTR is normally shorter.

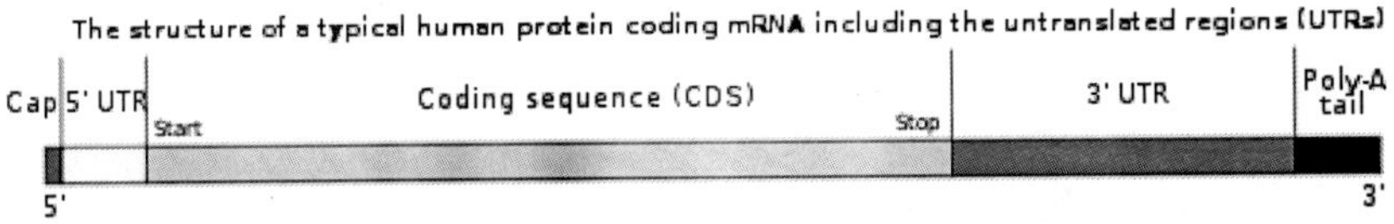

Figure: mRNA structure, approximately to scale for a human mRNA, where the median length of the 5'UTR is 170 nucleotides.

Several regulatory sequences may be found in the 5' UTR:

- Binding sites for proteins, that may affect the mRNA's stability or translation, for example iron responsive elements, that regulate gene expression in response to iron.
- Riboswitches.
- Sequences that promote or inhibit translation initiation.
- Introns within 5' UTRs have been linked to regulation of gene expression and mRNA export.

3' UTR

In molecular genetics, the three prime untranslated region (3' UTR) is a particular section of messenger RNA (mRNA). It follows the coding region. An mRNA molecule codes for a protein through translation. The mRNA also contains regions that are not translated. In eukaryotes these regions are the cap, 5' untranslated region, 3' untranslated region, and polyA tail.

In prokaryotes mRNA structures have some differences (see mRNA) as do histone mRNAs. However, both have 3' UTRs.

Several regulatory sequences are found in the 3' UTR:

- A polyadenylation signal, usually AAUAAA, or a slight variant. This marks the site of cleavage of the transcript approximately 30 base pairs past the signal, followed by the addition of several hundred adenine residues (poly-A tail).
- Binding sites for proteins, that may affect the mRNA's stability or location in the cell, like SECIS elements (which direct the ribosome to translate the codon UGA as selenocysteines rather than as a stop codon), or AU-rich elements (AREs), stretches consisting of mainly adenine and uracil nucleotides (which can either stabilize or destabilize the mRNA depending on the protein bound to it).
- Binding sites for miRNAs, a type of RNAi.

Poly(A) Tail

The 3' poly(A) tail is a long sequence of adenine nucleotides (often several hundred) added to the 3' end of the pre-mRNA. This tail promotes export from the nucleus and translation, and protects the mRNA from degradation.

Monocistronic Versus Polycistronic mRNA

An mRNA molecule is said to be monocistronic when it contains the genetic information to translate only a single protein chain (polypeptide). This is the case for most of the eukaryotic mRNAs. On the other hand, polycistronic mRNA carries several open reading frames (ORFs), each of which is translated into a polypeptide. These polypeptides usually have a related function (they often are the subunits composing a final complex protein) and their coding sequence is grouped and regulated together in an operon.

Most of the mRNA found in bacteria and archea is polycistronic. Dicistronic or bicistronic is the term used to describe an mRNA that encodes only two proteins.

mRNA Circularization

In eukaryotes it is thought that mRNA molecules form circular structures due to an interaction between the cap binding complex and poly(A)-binding protein. Circularization is thought to promote recycling of ribosomes on the same message leading to efficient translation.

Degradation

Different mRNAs within the same cell have distinct lifetimes (stabilities). In bacterial cells, individual mRNAs can survive from seconds to more than an hour; in mammalian cells, mRNA lifetimes range from several minutes to days.

The greater the stability of an mRNA the more protein may be produced from that mRNA. The limited lifetime of mRNA enables a cell to alter protein synthesis rapidly in response to its changing needs. There are many mechanisms that lead to the destruction of a mRNA, some of which are described below.

Prokaryotic mRNA Degradation

In general, in prokaryotes the lifetime of mRNA is much shorter than in eukaryotes. Prokaryotes degrade messages by using a combination of ribonucleases, including endonucleases, 3' exonucleases, and 5' exonucleases. In some instances, small RNA molecules (sRNA) tens to hundreds of nucleotides long can stimulate the degradation of specific mRNAs by base-pairing with complementary sequences and facilitating ribonuclease cleavage.

It was recently shown that bacteria also have a sort of 5' cap consisting of a triphosphate on the 5' end. Removal of two of the phosphates leaves a 5' monophosphate, causing the message to be destroyed by the endonuclease RNase E.

Eukaryotic mRNA Turnover

Inside eukaryotic cells, there is a balance between the processes of translation and mRNA decay. Messages that are being actively translated are bound by ribosomes, the eukaryotic initiation factors eIF-4E and eIF-4G, and poly(A)-binding protein. eIF-4E and eIF-4G block the decapping enzyme (DCP2), and poly(A)-binding protein blocks the exosome complex, protecting the ends of the message.

The balance between translation and decay is reflected in the size and abundance of cytoplasmic structures known as P-bodies The poly(A) tail of the mRNA is shortened by specialized exonucleases that are targeted to specific messenger RNAs by a combination of cis-regulatory sequences on the RNA and trans-acting RNA-binding proteins. Poly(A) tail removal is thought to disrupt the circular structure of the message and destabilize the cap binding complex.

The message is then subject to degradation by either the exosome complex or the decapping complex. In this way, translationally inactive messages can be destroyed quickly, while active messages remain intact. The mechanism by which translation stops and the message is handed-off to decay complexes is not understood in detail.

AU-rich Element Decay

The presence of AU-rich elements in some mammalian mRNAs tends to destabilize those transcripts through the action of cellular proteins that bind these sequences and stimulate poly(A) tail removal. Loss of the poly(A) tail is thought to promote mRNA degradation by facilitating attack by both the exosome complex and the decapping complex. Rapid mRNA degradation via AU-rich elements is a critical mechanism for preventing the overproduction of potent cytokines such as tumor necrosis factor (TNF) and granulocyte-macrophage colony stimulating factor (GM-CSF). AU-rich elements also regulate the biosynthesis of proto-oncogenic transcription factors like c-Jun and c-Fos.

Nonsense Mediated Decay

Nonsense-mediated decay (NMD) is a cellular mechanism of mRNA surveillance that functions to detect nonsense mutations and prevent the expression of truncated or erroneous proteins. Following transcription, precursor mRNA undergoes an assemblage of ribonucleoprotein (RNP) components followed by regulatory pre-mRNA processing. Large average intron size in eukaryotic cells greatly increases the probability that aberrant mRNA splicing will result in the presence of a nonsense (stop) codon (UAA, UAG, UGA) somewhere within the open reading frame.

NMD is triggered by exon junction complexes (EJCs) (components of the assembled RNP) that are deposited during pre-mRNA processing. While EJCs located in an open reading frame upstream of exon-exon junctions may facilitate ribosomal recruitment prior to displacement by a "pioneer" round of translation, EJCs located downstream of a nonsense codon are not displaced because the ribosome is released from the transcript before reaching it.

These remaining EJCs function as tags for recruitment of UPF1 following the mRNA's transport out of the nucleus and into the cytosol where the RNA is degraded, for example by the exosome complex. NMD is not only a mechanism for degrading aberrant mRNA, however, as there are numerous examples of normal transcripts whose expression is regulated by this process including the plasticity protein Arc/Arg3.1.

Eukaryotic messages are subject to surveillance by nonsense mediated decay (NMD), which checks for the presence of premature stop codons (nonsense codons) in the message. These can arise via incomplete splicing, V(D)J recombination in the adaptive immune system, mutations in DNA, transcription errors, leaky scanning by the ribosome causing a frame shift, and other causes. Detection of a premature stop codon triggers mRNA degradation by 5' decapping, 3' poly(A) tail removal, or endonucleolytic cleavage.

Small Interfering RNA (siRNA)

Small interfering RNA (siRNA), sometimes known as short interfering RNA or silencing RNA, is a class of double-stranded RNA molecules, 20-25 nucleotides in length, that play a variety of roles in biology.

The most notable role of siRNA is its involvement in the RNA interference (RNAi) pathway, where it interferes with the expression of a specific gene. In addition to its role in the RNAi pathway, siRNA also acts in RNAi-related pathways, e.g., as an antiviral mechanism or in shaping the chromatin structure of a genome; the complexity of these pathways is only now being elucidated.siRNAs were first discovered by David Baulcombe's group at the Sainsbury Laboratory in Norwich, England, as part of post-transcriptional gene silencing (PTGS) in plants.

The group published their findings in Science in a 1999 paper titled "A species of small antisense RNA in posttranscriptional gene silencing in plants".

Shortly thereafter, in 2001, synthetic siRNAs were shown to be able to induce RNAi in mammalian cells by Thomas Tuschl, and colleagues in a paper published in Nature. This discovery led to a surge in interest in harnessing RNAi for biomedical research and drug development.

In metazoans, small interfering RNAs (siRNAs) processed by Dicer are incorporated into a complex known as the RNA-induced silencing complex or RISC. This complex contains an endonuclease that cleaves perfectly complementary messages to which the siRNA binds.

The resulting mRNA fragments are then destroyed by exonucleases. siRNA is commonly used in laboratories to block the function of genes in cell culture.

It is thought to be part of the innate immune system as a defense against double-stranded RNA viruses. Structure siRNAs have a well-defined structure: a short (usually 21-nt) double-strand RNA (dsRNA) with 2-nt 3' overhangs on either end:

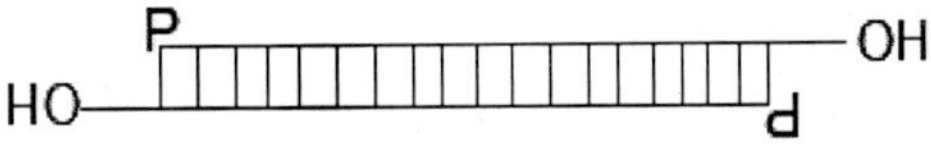

Schematic representation of a siRNA molecule: a ~19-21basepair RNA core duplex that is followed by a 2 nucleotide 3' overhang on each strand. OH: 3' hydroxyl; P: 5' phosphate.

Each strand has a 5' phosphate group and a 3' hydroxyl (-OH) group. This structure is the result of processing by dicer, an enzyme that converts either long dsRNAs or small hairpin RNAs into siRNAs. siRNAs can also be exogenously (artificially) introduced into cells by various transfection methods to bring about the specific knockdown of a gene of interest. In essence, any gene whose sequence is known can, thus, be targeted based on sequence complementarity with an appropriately tailored siRNA. This has made siRNAs an important tool for gene function and drug target validation studies in the post-genomic era.

RNAi Induction Using siRNAs or Their Biosynthetic Precursors

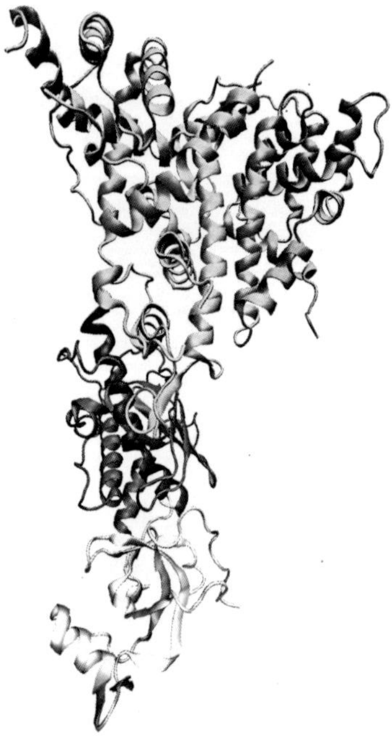

Figure: *Dicer protein colored by protein domain.*

Transfection of an exogenous siRNA can be problematic because the gene knockdown effect is only transient, in particular, in rapidly dividing cells. One way of overcoming this challenge is to modify the siRNA in such a way as to allow it to be expressed by an appropriate vector, e.g., a plasmid.

This is done by the introduction of a loop between the two strands, thus producing a single transcript, which can be processed into a functional siRNA . Such transcription cassettes typically use an RNA polymerase III promoter (e.g., U6 or H1), which usually directs the transcription of small nuclear RNAs (snRNAs) (U6 is involved in gene splicing; H1 is the RNase component of human RNase P). It is assumed (although not known for certain) that the resulting siRNA transcript is then processed by Dicer.

RNA Activation

It has recently been found that dsRNA can also activate gene expression, a mechanism that has been termed "small RNA-induced gene activation" or RNAa. It has been shown that dsRNAs targeting gene promoters induce potent transcriptional activation of associated genes. RNAa was demonstrated in human cells using synthetic dsRNAs, termed "small activating RNAs" (saRNAs). It is currently not known whether RNAa is conserved in other organisms.

Challenges: Avoiding Nonspecific Effects

Because RNAi intersects with a number of other pathways, it is not surprising that on occasion nonspecific effects are triggered by the experimental introduction of an siRNA. When a mammalian cell encounters a double-stranded RNA such as an siRNA, it may mistake it as a viral by-product and mount an immune response. Furthermore, because structurally related microRNAs modulate gene expression largely via incomplete complementarity base pair interactions with a target mRNA, the introduction of an siRNA may cause unintended off-targeting.

Innate Immunity

Introduction of too much siRNA can result in nonspecific events due to activation of innate immune responses. Most evidence to date suggests that this is probably due to activation of the dsRNA sensor PKR, although retinoic acid-inducible gene I (RIG-I) may also be involved .

The induction of cytokines via toll-like receptor 7 (TLR7) has also been described. One promising method of reducing the nonspecific effects is to convert the siRNA into a microRNA. MicroRNAs occur naturally, and by harnessing this endogenous pathway it should be possible to achieve similar gene knockdown at comparatively low concentrations of resulting siRNAs. This should minimize nonspecific effects.

Off-targeting

Off-targeting is another challenge to the use of siRNAs as a gene knockdown tool. Here, genes with incomplete complementarity are inadvertently downregulated by the siRNA (in effect, the siRNA acts as a miRNA), leading to problems in data interpretation and potential toxicity. This, however, can be partly addressed by designing appropriate control experiments, and siRNA design algorithms are currently being developed to produce siRNAs free from off-targeting. Genome-wide expression analysis, e.g., by microarray technology, can then be used to verify this and further refine the algorithms. A 2006 paper from the laboratory of Dr. Khvorova implicates 6- or 7-basepair-long stretches from position 2 onward in the siRNA matching with 3'UTR regions in off-targeted genes.

Possible Therapeutic Applications and Challenges

Given the ability to knock down, in essence, any gene of interest, RNAi via siRNAs has generated a great deal of interest in both basic and applied biology. There are an increasing number of large-scale RNAi screens that are designed to identify the important genes in various biological pathways. Because disease processes also depend on the activity of multiple genes, it is expected that in some situations turning off the activity of a gene with an siRNA could produce a therapeutic benefit.

However, applying RNAi via siRNAs to living animals, especially humans, poses many challenges. Under experiments, siRNAs show different effectiveness in different cell types in a manner as yet poorly understood: Some cells respond well to siRNAs and show a robust knockdown, whereas others show no such knockdown (even despite efficient transfection).

Phase I results of the first two therapeutic RNAi trials (indicated for age-related macular degeneration, aka AMD) reported at the end of 2005 that siRNAs are well tolerated and have suitable pharmacokinetic properties. Proof of concept trials have indicated that Ebola-targeted siRNAs may be effective as post-exposure prophylaxis in humans, with 100% of non-human primates surviving a lethal dose of Zaire Ebolavirus, the most lethal strain.

MicroRNA (miRNA)

A microRNA (abbreviated miRNA) is a short ribonucleic acid (RNA) molecule found in eukaryotic cells. A microRNA molecule has very few nucleotides (an average of 22) compared with other RNAs.

miRNAs are post-transcriptional regulators that bind to complementary sequences on target messenger RNA transcripts (mRNAs), usually resulting in translational repression or target degradation and gene silencing. The human genome may encode over 1000 miRNAs, which may target about 60% of mammalian genes and are abundant in many human cell types. miRNAs show very different characteristics between plants and metazoans. In plants the miRNA complementarity to its mRNA target is nearly perfect, with no or few mismatched bases. In metazoans, on the other hand, miRNA complementarity typically encompasses the 5' bases 2-7 of the microRNA, the microRNA seed region, and one miRNA can target many different sites on the same mRNA or on many different mRNAs. Another difference is the location of target sites on mRNAs. In metazoans, the miRNA target sites are in the three prime untranslated regions (3'UTR) of the mRNA. In plants, targets can be located in the 3' UTR but are more often in the coding region itself. MiRNAs are well conserved in eukaryotic organisms and are thought to be a vital and evolutionarily ancient component of genetic regulation.

The first miRNAs were characterized in the early 1990s. However, miRNAs were not recognized as a distinct class of biological regulators with conserved functions until the early 2000s. Since then, miRNA research has revealed multiple roles in negative regulation (transcript degradation and sequestering, translational suppression) and possible involvement in positive regulation (transcriptional and translational activation). By affecting gene regulation, miRNAs are likely to be involved in most biological processes. Different sets of expressed miRNAs are found in different cell types and tissues. Aberrant expression of miRNAs has been implicated in numerous disease states, and miRNA-based therapies are under investigation.

MicroRNAs (miRNAs) are small RNAs that typically are partially complementary to sequences in metazoan messenger RNAs. Binding of a miRNA to a message can repress translation of that message and accelerate poly(A) tail removal, thereby hastening mRNA degradation. The mechanism of action of miRNAs is the subject of active research.

History

MicroRNAs were discovered in 1993 by Victor Ambros, Rosalind Lee and Rhonda Feinbaum during a study of the gene lin-14 in C. elegans development. They found that LIN-14 protein abundance was regulated by a short RNA product encoded by the lin-4 gene. A 61-nucleotide precursor from the lin-4 gene matured to a 22-nucleotide RNA that contained sequences partially complementary to multiple sequences in the 3' UTR of the lin-14 mRNA.

This complementarity was both necessary and sufficient to inhibit the translation of the lin-14 mRNA into the LIN-14 protein. Retrospectively, the lin-4 small RNA was the first microRNA to be identified, though at the time, it was thought to be a nematode idiosyncrasy. Only in 2000 was a second RNA characterized: let-7, which repressed lin-41, lin-14, lin-28, lin-42, and daf-12 expression during developmental stage transitions in C. elegans. let-7 was soon found to be conserved in many species, indicating the existence of a wider phenomenon.

Nomenclature

Under a standard nomenclature system, names are assigned to experimentally confirmed miRNAs before publication of their discovery. The prefix "mir" is followed by a dash and a number, the latter often indicating order of naming. For example, mir-123 was named and likely discovered prior to mir-456.

The uncapitalized "mir-" refers to the pre-miRNA, while a capitalized "miR-" refers to the mature form. miRNAs with nearly identical sequences bar one or two nucleotides are annotated with an additional lower case letter. For example, miR-123a would be closely related to miR-123b.

Pre-miRNAs that lead to 100% identical mature miRNAs but that are located at different places in the genome are indicated with an additional dash-number suffix. For example, the pre-miRNAs hsa-mir-194-1 and hsa-mir-194-2 lead to an identical mature miRNA (hsa-miR-194) but are located in different regions of the genome. Species of origin is designated with a three-letter prefix, e.g., hsa-miR-123 is a human (Homo sapiens)miRNA and oar-miR-123 is a sheep (Ovis aries) miRNA.

Other common prefixes include 'v' for viral (miRNA encoded by a viral genome) and 'd' for Drosophila miRNA (a fruit fly commonly studied in genetic research). When two mature microRNAs originate from opposite arms of the same pre-miRNA, they are denoted with a -3p or -5p suffix. (In the past, this distinction was also made with 's' (sense) and 'as' (antisense)). When relative expression levels are known, an asterisk following the name indicates an miRNA expressed at low levels relative to the miRNA in the opposite arm of a hairpin. For example, miR-123 and miR-123* would share a pre-miRNA hairpin, but more miR-123 would be found in the cell.

Biogenesis

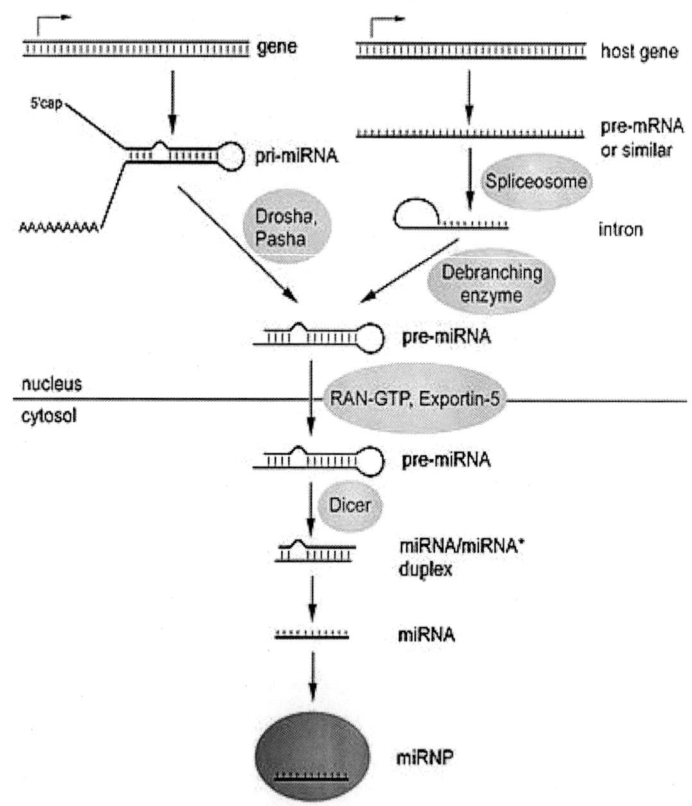

Figure: MicroRNAs are produced from either their own genes or from introns. A video of this process can be found here.

The majority of the characterized miRNA genes are intergenic or oriented antisense to neighbouring genes and are therefore suspected to be transcribed as independent units As much as 40% of miRNA genes may lie in the introns of protein and non-protein coding genes or even in exons of long nonprotein-coding transcripts. These are usually, though not exclusively, found in a sense orientation. and thus usually are regulated together with their host genes. Other miRNA genes showing a common promoter include the 42-48% of all miRNAs originating from polycistronic units containing multiple discrete loops from which mature miRNAs are processed, although this does not necessarily mean the mature miRNAs of a family will be homologous in structure and function.

The promoters mentioned have been shown to have some similarities in their motifs to promoters of other genes transcribed by RNA polymerase II such as protein coding genes. The DNA template is not the final word on mature miRNA production: 6% of human miRNAs show RNA editing (IsomiRs), the site-specific modification of RNA sequences to yield products different from those encoded by their DNA. This increases the diversity and scope of miRNA action beyond that implicated from the genome alone.

Transcription

miRNA genes are usually transcribed by RNA polymerase II (Pol II). The polymerase often binds to a promoter found near the DNA sequence encoding what will become the hairpin loop of the pre-miRNA. The resulting transcript is capped with a specially-modified nucleotide at the 5' end, polyadenylated with multiple adenosines (a poly(A) tail), and spliced. Animal miRNAs are initially transcribed as part of one arm of an <"80 nucleotide RNA stem-loop that in turn forms part of a several hundred nucleotides long miRNA precursor termed a primary miRNA (pri-miRNA)s. When a stem-loop precursor is found in the 3' UTR, a transcript may serve as a pri-miRNA and a mRNA. RNA polymerase III (Pol III) transcribes some miRNAs, especially those with upstream Alu sequences, transfer RNAs (tRNAs), and mammalian wide interspersed repeat (MWIR) promoter units.

Nuclear Processing

A single pri-miRNA may contain from one to six miRNA precursors. These hairpin loop structures are composed of about 70 nucleotides each. Each hairpin is flanked by sequences necessary for efficient processing. The double-stranded RNA structure of the hairpins in a pri-miRNA is recognized by a nuclear protein known as DiGeorge Syndrome Critical Region 8 (DGCR8 or "Pasha" in invertebrates), named for its association with DiGeorge Syndrome. DGCR8 associates with the enzyme Drosha, a protein that cuts RNA, to form the "Microprocessor" complex. In this complex, DGCR8 orients the catalytic RNase III domain of Drosha to liberate hairpins from pri-miRNAs by cleaving RNA about eleven nucleotides from the hairpin base (two helical RNA turns into the stem). The product resulting has a two-nucleotide overhang at its 3' end; it has 3' hydroxyl and 5' phosphate groups. It is often termed as a pre-miRNA (precursor-miRNA). pre-miRNAs that are spliced directly out of introns, bypassing the Microprocessor complex, are known as "Mirtrons." Originally thought to exist only in Drosophila and C. elegans, mirtrons have now been found in mammals. Perhaps as many as 16% of pri-miRNAs may be altered through nuclear RNA editing. Most commonly, enzymes known as adenosine deaminases acting on RNA (ADARs) catalyze adenosine to inosine (A to I) transitions. RNA editing can halt nuclear processing (for example, of pri-miR-142, leading to degradation by the ribonuclease Tudor-SN) and alter downstream processes including cytoplasmic miRNA processing and target specificity (e.g., by changing the seed region of miR-376 in the central nervous system).

Nuclear Export

pre-miRNA hairpins are exported from the nucleus in a process involving the nucleocytoplasmic shuttle Exportin-5. This protein, a member of the karyopherin family, recognizes a two-nucleotide overhang left by the RNase III enzyme Drosha at the 3' end of the pre-miRNA hairpin. Exportin-5-mediated transport to the cytoplasm is energy-dependent, using GTP bound to the Ran protein.

Cytoplasmic Processing

In cytoplasm, the pre-miRNA hairpin is cleaved by the RNase III enzyme Dicer. This endoribonuclease interacts with the 3' end of the hairpin and cuts away the loop joining the 3' and 5' arms, yielding an imperfect miRNA:miRNA* duplex about 22 nucleotides in length. Overall hairpin length and loop size influence the efficiency of Dicer processing, and the imperfect nature of the miRNA:miRNA* pairing also affects cleavage.

Although either strand of the duplex may potentially act as a functional miRNA, only one strand is usually incorporated into the RNA-induced silencing complex (RISC) where the miRNA and its mRNA target interact.

Biogenesis in Plants

miRNA biogenesis in plants differs from metazoan biogenesis mainly in the steps of nuclear processing and export. Instead of being cleaved by two different enzymes, once inside and once outside the nucleus, both cleavages of the plant miRNA is performed by a Dicer homolog, called Dicer-like1 (DL1). DL1 is only expressed in the nucleus of plant cells, which indicates that both reactions take place inside the nucleus.

Before plant miRNA:miRNA* duplexes are transported out of the nucleus its 3' overhangs are methylated by a RNA methyltransferaseprotein called

Hua-Enhancer1 (HEN1). The duplex is then transported out of the nucleus to the cytoplasm by a protein called Hasty (HST), an Exportin 5 homolog, where they disassemble and the mature miRNA is incorporated into the RISC.

The RNA-induced Silencing Complex

The mature miRNA is part of an active RNA-induced silencing complex (RISC) containing Dicer and many associated proteins. RISC is also known as a microRNA ribonucleoprotein complex (miRNP); RISC with incorporated miRNA is sometimes referred to as "miRISC."

Dicer processing of the pre-miRNA is thought to be coupled with unwinding of the duplex. Generally, only one strand is incorporated into the miRISC, selected on the basis of its thermodynamic instability and weaker base-pairing relative to the other strand. The position of the stem-loop may also influence strand choice. The other strand, called the passenger strand due to its lower levels in the steady state, is denoted with an asterisk (*) and is normally degraded. In some cases, both strands of the duplex are viable and become functional miRNA that target different mRNA populations.

Members of the argonaute (Ago) protein family are central to RISC function. Argonautes are needed for miRNA-induced silencing and contain two conserved RNA binding domains: a PAZ domain that can bind the single stranded 3' end of the mature miRNA and a PIWI domain that structurally resembles ribonuclease-H and functions to interact with the 5' end of the guide strand.

They bind the mature miRNA and orient it for interaction with a target mRNA. Some argonautes, for example human Ago2, cleave target transcripts directly; argonautes may also recruit additional proteins to achieve translational repression. The human genome encodes eight argonaute proteins divided by sequence similarities into two families:

AGO (with four members present in all mammalian cells and called E1F2C/hAgo in humans), and PIWI (found in the germ line and hematopoietic stem cells).

Additional RISC components include TRBP [human immunodeficiency virus (HIV) transactivating response RNA (TAR) binding protein], PACT (protein activator of the interferon induced protein kinase (PACT), the SMN complex, fragile X mental retardation protein (FMRP), and Tudor staphylococcal nuclease-domain-containing protein (Tudor-SN). Strands

The RNA endonuclease Dicer plays a role in aiding RISC action by providing the initial RNA material to activate the complex as well as the first RNA substrate molecule. When the Dicer, which has endonuclease activity against dsRNA and pre-miRNAs, cleaves a pre-miRNA stem-loop or a dsRNA, a 20-

to 25-base-pair double-stranded RNA fragment is formed with a two-nucleotide 3' overhang at each end.

- One strand is integrated into the RISC complex. This strand is known as the guide strand and is selected by the argonaute protein, the catalytically active RNase in the RISC complex, on the basis of the thermodynamic stability of the 5' end. The strand with a less thermodynamically stable 5' end is loaded into the RISC complex.
- The remaining strand, known as the anti-guide or passenger strand, is degraded as a RISC complex substrate.

Binding

The RISC complex with a bound siRNA recognizes complementary messenger RNA (mRNA) molecules and degrades them, resulting in substantially decreased levels of protein translation and effectively turning off the gene.

It is as yet unclear how the activated RISC complex locates the mRNA targets in the cell, though it has been shown that the process can occur in situations outside of ongoing protein translation from mRNA. Endogenously expressed miRNA in metazoans is usually imperfectly complementary to a large number of nuclear genes and has a modulating effect on these genes' levels of expression via translational repression.

However, in plants the process has a much greater specificity to the mRNA and usually each miRNA only binds to one mRNA. Recently identified members of RISC complex are SND1(p100 Coactivator or Tudor SN)and AEG-1(MTDH). AEG-1 and SND1 are oncogenes and regulates various gene expression.

Mode of Silencing

Gene silencing may occur either via mRNA degradation or preventing mRNA from being translated. It has been demonstrated that if there is complete complementation between the miRNA and target mRNA sequence, Ago2 can cleave the mRNA and lead to direct mRNA degradation. Yet, if there isn't complete complementation the silencing is achieved by preventing translation.

miRNA Turnover

Turnover of mature miRNA is needed for rapid changes in miRNA expression profiles. During miRNA maturation in the cytoplasm, uptake by the Argonaute protein is thought to stabilize the guide strand, while the opposite (* or "passenger") strand is preferentially destroyed. In what has been called a "Use it or lose it" strategy, Argonaute may preferentially retain miRNAs with many targets over miRNAs with few or no targets, leading to degradation of the non-targeting molecules.

Decay of mature miRNAs in Caenorhabditis elegans is mediated by the 5′-to-3′ exoribonuclease XRN2, also known as Rat1p. In plants, SDN (small RNA degrading nuclease) family members degrade miRNAs in the opposite (3'-to-5') direction. Similar enzymes are encoded in animal genomes, but their roles have not yet been described.

Several miRNA modifications affect miRNA stability. As indicated by work in the model organism Arabidopsis thaliana (thale cress), mature plant miRNAs appear to be stabilized by the addition of methyl moieties at the 3' end. The 2'-O-conjugated methyl groups block the addition of uracil (U) residues by uridyltransferase enzymes, a modification that may be associated with miRNA degradation. However, uridylation may also protect some miRNAs; the consequences of this modification are incompletely understood. Uridylation of some animal miRNAs has also been reported. Both plant and animal miRNAs may be altered by addition of adenine (A) residues to the 3' end of the miRNA. An extra A added to the end of mammalian miR-122, a liver-enriched miRNA important in Hepatitis C, stabilizes the molecule, and plant miRNAs ending with an adenine residue have slower decay rates.

Cellular Functions

The function of miRNAs appears to be in gene regulation. For that purpose, a miRNA is complementary to a part of one or more messenger RNAs (mRNAs). Animal miRNAs are usually complementary to a site in the 3' UTR whereas plant miRNAs are usually complementary to coding regions of mRNAs. Perfect or near perfect base pairing with the target RNA promotes cleavage of the RNA. This is the primary mode of plant miRNAs. In animals miRNAs more often have only partly the right sequence of nucleotides to bond with the target mRNA. The match-ups are imperfect. Animal miRNAs inhibit protein translation of the target mRNA (this exists in plants as well but is less common). MicroRNAs that are partially complementary to a target can also speed up deadenylation, causing mRNAs to be degraded sooner. For partially complementary microRNAs to recognise their targets, nucleotides 2–7 of the miRNA (its 'seed region') still have to be perfectly complementary. miRNAs occasionally also cause histone modification and DNA methylation of promoter sites, which affects the expression of target genes.

Unlike plant microRNAs, the animal microRNAs target a diverse set of genes. However, genes involved in functions common to all cells, such as gene expression, have relatively fewer microRNA target sites and seem to be under selection to avoid targeting by microRNAs.

dsRNA can also activate gene expression, a mechanism that has been termed "small RNA-induced gene activation" or RNAa. dsRNAs targeting gene promoters can induce potent transcriptional activation of associated genes. This was demonstrated in human cells us as also been demonstrated for endogenous microRNA.

Interactions between microRNAs and complementary sequences on genes and even pseudogenes that share sequence homology are thought to be a back channel of communication regulating expression levels between paralogous genes. Given the name "competing endogenous RNAs" (ceRNAs), these microRNAs bind to "microRNA response elements" on genes and pseudogenes and may provide another explanation for the persistence of non-coding ("junk") DNA.

Evolution

MicroRNAs are significant phylogenetic markers because of their astonishingly low rate of evolution. Their origin may have permitted the development of morphological innovation, and by making gene expression more specific and 'fine-tunable', permitted the genesis of complex organs and perhaps, ultimately, complex life. Indeed, rapid bursts of morphological innovation are generally associated with a high rate of microRNA accumulation.

MicroRNAs originate predominantly by the random formation of hairpins in "non-coding" sections of DNA (i.e. introns or intergene regions), but also by the duplication and modification of existing microRNAs. The rate of evolution (i.e. nucleotide substitution) in recently-originated microRNAs is comparable to that elsewhere in the non-coding DNA, implying evolution by neutral drift; however, older microRNAs have a much lower rate of change (often less than one substitution per hundred million years), suggesting that once a microRNA gains a function it undergoes extreme purifying selection. At this point, a microRNA is rarely lost from an animal's genome, although microRNAs which are more recently derived (and thus presumably non-functional) are frequently lost. This makes them a valuable phylogenetic marker, and they are being looked upon as a possible solution to such outstanding phylogenetic problems as the relationships of arthropods.

MicroRNAs feature in the genomes of most eukaryotic organisms, from the brown algae to the metazoa. Across all species, in excess of 5000 had been identified by March 2010. Whilst short RNA sequences (50 – hundreds of base pairs) of a broadly comparable function occur in bacteria, bacteria lack true microRNAs.

Experimental Detection and Manipulation of miRNA

MicroRNA expression can be quantified in a two-step polymerase chain reaction process of modified RT-PCR followed by quantitative real-time PCR.

Variations of this method achieve absolute or relative quantification. miRNAs can also be hybridized to microarrays, slides or chips with probes to hundreds or thousands of miRNA targets, so that relative levels of miRNAs can be determined in different samples. MicroRNAs can be both discovered and profiled by high-throughput sequencing methods.

The activity of an miRNA can be experimentally inhibited using a locked nucleic acid (LNA) oligo, a Morpholino oligo or a 2'-O-methyl RNA oligo. Additionally, a specific miRNA can be silenced by a complementary antagomir.

MicroRNA maturation can be inhibited at several points by steric-blocking oligos. The miRNA target site of an mRNA transcript can also be blocked by a steric-blocking oligo. For the "in situ" detection of miRNA, the use of LNA is currently the only efficient method. The locked conformation of LNA results in enhanced hybridization properties and increases sensitivity and selectivity, making it ideal for detection of short miRNA.

miRNA and Disease

Just as miRNA is involved in the normal functioning of eukaryotic cells, so has dysregulation of miRNA been associated with disease. A manually curated, publicly available database miR2Disease documents known relationships between miRNA dysregulation and human disease.

miRNA and Inherited Diseases

A mutation in the seed region of miR-96 causes hereditary progressive hearing loss. A mutation in the seed region of miR-184 causes hereditary keratoconus with anterior polar cataract. Deletion of the miR-17~92 cluster causes skeletal and growth defects.

miRNA and Cancer

Several miRNAs have been found to have links with some types of cancer. MicroRNA-21 is one of the first microRNAs that was identified as an oncomiR.

A study of mice altered to produce excess c-Myc — a protein with mutated forms implicated in several cancers — shows that miRNA has an effect on the development of cancer. Mice that were engineered to produce a surplus of types of miRNA found in lymphoma cells developed the disease within 50 days and died two weeks later. In contrast, mice without the surplus miRNA lived over 100 days. Leukemia can be caused by the insertion of a viral genome next to the 17-92 array of microRNAs leading to increased expression of this microRNA.

Another study found that two types of miRNA inhibit the E2F1 protein, which regulates cell proliferation. miRNA appears to bind to messenger RNA before it can be translated to proteins that switch genes on and off. By measuring activity among 217 genes encoding miRNA, patterns of gene activity that can distinguish types of cancers can be discerned. miRNA signatures may enable classification of cancer. This will allow doctors to determine the original tissue type which spawned a cancer and to be able to target a treatment course based on the original tissue type. miRNA profiling has already been able to determine whether patients with chronic lymphocytic leukemia had slow growing or aggressive forms of the cancer.

Transgenic mice that over-express or lack specific miRNAs have provided insight into the role of small RNAs in various malignancies.

A novel miRNA-profiling based screening assay for the detection of early-stage colorectal cancer has been developed and is currently in clinical trials. Early results showed that blood plasma samples collected from patients with early, resectable (Stage II) colorectal cancer could be distinguished from those of sex-and age-matched healthy volunteers. Sufficient selectivity and specificity could be achieved using small (less than 1 mL) samples of blood. The test has potential to be a cost-effective, non-invasive way to identify at-risk patients who should undergo colonoscopy.

Another role for miRNA in cancers is to use their expression level as a prognostic, for example one study on NSCLC samples found that low miR-324a levels could serve as a prognostic indicator of poor survival, another found that either high miR-185 or low miR-133b levels correlated with metastasis and poor survival in colorectal cancer.

miRNA and Heart Disease

The global role of miRNA function in the heart has been addressed by conditionally inhibiting miRNA maturation in the murine heart, and has revealed that miRNAs play an essential role during its development. miRNA expression profiling studies demonstrate that expression levels of specific miRNAs change in diseased human hearts, pointing to their involvement in cardiomyopathies. Furthermore, studies on specific miRNAs in animal models have identified distinct roles for miRNAs both during heart development and under pathological conditions, including the regulation of key factors important for cardiogenesis, the hypertrophic growth response, and cardiac conductance.

miRNA and the Nervous System

miRNAs appear to regulate the nervous system. Neural miRNAs are involved at various stages of synaptic development, including dendritogenesis (involving miR-132, miR-134 and miR-124), synapse formation and synapse maturation (where miR-134 and miR-138 are thought to be involved). Some studies find altered miRNA expression in schizophrenia.

miRNA and non-coding RNAs

When the human genome project mapped its first chromosome in 1999, it was predicted the genome would contain over 100,000 protein coding genes. However, only around 20,000 were eventually identified (International Human Genome Sequencing Consortium, 2004).

Since then, the advent of bioinformatics approaches combined with genome tiling studies examining the transcriptome, systematic sequencing of full length cDNA libraries, and experimental validation (including the creation of miRNA derived antisense oligonucleotides called antagomirs) have revealed that many transcripts are non protein-coding RNA, including several snoRNAs and miRNAs.

Other Decay Mechanisms

There are other ways by which messages can be degraded, including non-stop decay and silencing by Piwi-interacting RNA (piRNA), among others.

Untranslated region (UTR)

In molecular genetics, an untranslated region (or UTR) refers to either of two sections on each side of a coding sequence on a strand of mRNA. If it is found on the 5' side, it is called the 5' UTR (or leader sequence), or if it is found on the 3' side, it is called the 3' UTR (or trailer sequence).

Primary transcript

A primary transcript is an RNA molecule that has not yet undergone any modification after its synthesis. For example, a precursor messenger RNA (pre-mRNA) is a primary transcript that becomes a messenger RNA (mRNA) after processing, and a primary microRNA (pri-miRNA) precursor becomes a microRNA (miRNA) after processing.

Spliceosome

A spliceosome is a complex of snRNA and protein subunits that removes introns from a transcribed pre-mRNA (hnRNA) segment. This process is generally referred to as splicing.

Composition

Each spliceosome is composed of five small nuclear RNAs (snRNAs), and a range of associated protein factors. The snRNAs that make up the nuclear spliceosome are named U1, U2, U4, U5, and U6, and participate in several RNA-RNA and RNA-protein interactions. The RNA component of the small nuclear ribonucleic protein or snRNP (pronounced "snurp") is rich in uridine (the nucleoside analog of the uracil nucleotide). The canonical assembly of the spliceosome occurs anew on each hnRNA (pre-mRNA). The hnRNA contains specific sequence elements that are recognized and utilized during spliceosome assembly. These include the 5' end splice, the branch point sequence, the polypyrimidine tract, and the 3' end splice site. The spliceosome catalyzes the removal of introns, and the ligation of the flanking exons. Introns typically have a GU nucleotide sequence at the 5' end splice site, and an AG at the 3' end splice site.

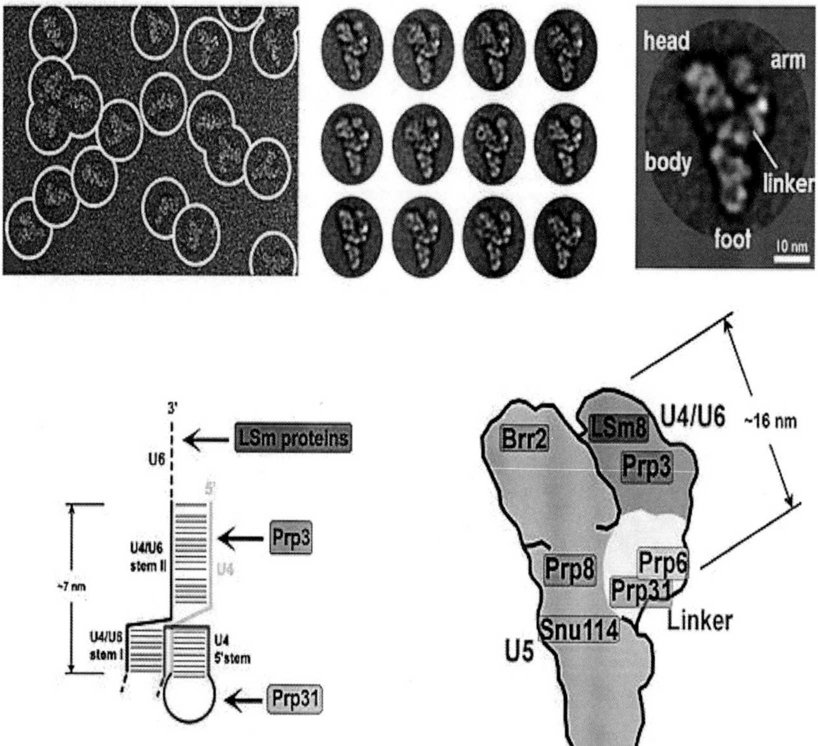

Figure: *Above are electron microscopy fields of negatively stained yeast (Saccharomyces cerevisiae) tri-snRNPs. Below left is a schematic illustration of the interaction of tri-snRNP proteins with the U4/U6 snRNA duplex. Below right is a cartoon model of the yeast tri-snRNP with shaded areas corresponding to U5 (gray), U4/U6 (orange) and the linker region (yellow).*

The 3' splice site can be further defined by a variable length of polypyrimidines, called the polypyrimidine tract (PPT), which serves the dual function of recruiting factors to the 3' splice site and possibly recruiting factors to the branch point sequence (BPS). The BPS contains the conserved Adenosine required for the first step of splicing. A group of less abundant snRNAs, U11, U12, U4atac, and U6atac, together with U5, are subunits of the so-called minor spliceosome that splices a rare class of pre-mRNA introns, denoted U12-type.

These snRNPs form the U12 spliceosome are located in the cytosol. New evidence derived from the first crystal structure of a group II intron suggests that the spliceosome is actually a ribozyme, and that it uses a two–metal ion mechanism for catalysis.

Alternative Splicing

Alternative splicing (the re-combination of different exons) is a major source of genetic diversity in eukaryotes. Splice variants have been used to account for the relatively small number of genes in the human genome. For years the estimate widely varied with top estimates reaching 100,000 genes, but now, due to the Human Genome Project the figure is believed to be closer to 20,000 genes. One particular Drosophila gene (Dscam, the Drosophila homolog of the human Down syndrome cell adhesion molecule DSCAM) can be alternatively spliced into 38,000 different mRNA.

RNA Splicing

In molecular biology and genetics, splicing is a modification of an RNA after transcription, in which introns are removed and exons are joined. This is needed for the typical eukaryotic messenger RNA before it can be used to produce a correct protein through translation. For many eukaryotic introns, splicing is done in a series of reactions which are catalyzed by the spliceosome, a complex of small nuclear ribonucleoproteins (snRNPs), but there are also self-splicing introns.

In 1977, work by the Sharp and Roberts labs revealed that genes of higher organisms are "split" or present in several distinct segments along the DNA molecule. The coding regions of the gene are separated by non-coding DNA that is not involved in protein expression. The split gene structure was found when adenoviral mRNAs were hybridized to endonuclease cleavage fragments of single stranded viral DNA.

It was observed that the mRNAs of the mRNA-DNA hybrids contained 5' and 3' tails of non-hydrogen bonded regions. When larger fragments of viral DNAs were used, forked structures of looped out DNA were observed when hybridized to the viral mRNAs. It was realized that the looped out regions, the introns, are excised from the precursor mRNAs in a process Sharp named "splicing". The split gene structure was subsequently found to be common to most eukaryotic genes. Phillip Sharp and Richard J. Roberts were awarded the 1993 Nobel Prize in Physiology or Medicine for their discovery of introns and the splicing process.

Spliceosome Assembly

The model for formation of the spliceosome active site involves an ordered, stepwise assembly of discrete snRNP particles on the hnRNA substrate. The first recognition of hnRNAs involves U1 snRNP binding to the 5' end splice site of the hnRNA and other non-snRNP associated factors to form the commitment complex, or early (E) complex in mammals. The commitment complex is an ATP-independent complex that commits the hnRNA to the splicing pathway.

U2 snRNP is recruited to the branch region through interactions with the E complex component U2AF (U2 snRNP auxiliary factor) and possibly U1 snRNP. In an ATP-dependent reaction, U2 snRNP becomes tightly associated with the branch point sequence (BPS) to form complex A. A duplex formed between U2 snRNP and the hnRNA branch region bulges out the branch adenosine specifying it as the nucleophile for the first transesterification.

The presence of a pseudouridine residue in U2 snRNA, nearly opposite of the branch site, results in an altered conformation of the RNA-RNA duplex upon the U2 snRNP binding. Specifically, the altered structure of the duplex induced by the pseudouridine places the 2' OH of the bulged adenosine in a favorable position for the first step of splicing. The U4/U5/U6 tri-snRNP is recruited to the assembling spliceosome to form complex B, and following several rearrangements, complex C (the spliceosome) is activated for catalysis. It is unclear how the triple snRNP is recruited to complex A, but this process may be mediated through protein-protein interactions and/or base pairing interactions between U2 snRNA and U6 snRNA.

The U5 snRNP interacts with sequences at the 5' and 3' splice sites via the invariant loop of U5 snRNA and U5 protein components interact with the 3' splice site region.

Upon recruitment of the triple snRNP, several RNA-RNA rearrangements precede the first catalytic step and further rearrangements occur in the catalytically active spliceosome. Several of the RNA-RNA interactions are mutually exclusive; however, it is not known what triggers these interactions, nor the order of these rearrangements.

The first rearrangement is probably the displacement of U1 snRNP from the 5' splice site and formation of a U6 snRNA interaction. It is known that U1 snRNP is only weakly associated with fully formed spliceosomes, and U1 snRNP is inhibitory to the formation of a U6-5' splice site interaction on a model of substrate oligonucleotide containing a short 5' exon and 5' splice site. Binding of U2 snRNP to the branch point sequence (BPS) is one example of an RNA-RNA interaction displacing a protein-RNA interaction. Upon recruitment of U2 snRNP, the branch binding protein SF1 in the commitment complex is displaced since the binding site of U2 snRNA and SF1 are mutually exclusive events.

Within the U2 snRNA, there are other mutually exclusive rearrangements that occur between competing conformations. For example, in the active form, stem loop IIa is favored; in the inactive form a mutually exclusive interaction between the loop and a downstream sequence predominates. It is unclear how U4 is displaced from U6 snRNAm, although RNA has been implicated in spliceosome assembly, and may function to unwind U4/U6 and promote the formation of a U2/U6 snRNA interaction.

The interactions of U4/U6 stem loops I and II dissociate and the freed stem loop II region of U6 folds on itself to form an intramolecular stem loop and U4 is no longer required in further spliceosome assembly.

The freed stem loop I region of U6 base pairs with U2 snRNA forming the U2/U6 helix I. However, the helix I structure is mutually exclusive with the 3' half of an internal 5' stem loop region of U2 snRNA.

Splicing Pathways

Several methods of RNA splicing occur in nature; the type of splicing depends on the structure of the spliced intron and the catalysts required for splicing to occur.

Spliceosomal Introns

Spliceosomal introns often reside in between eukaryotic protein-coding genes. Within the intron, a 3' splice site, 5' splice site, and branch site are required for splicing. The 5' splice site or splice donor site includes an almost invariant sequence GU at the 5' end of the intron, within a larger, less highly conserved consensus region. The 3' splice site or splice acceptor site terminates the intron with an almost invariant AG sequence.

Upstream (5'-ward) from the AG there is a region high in pyrimidines (C and U), or polypyrimidine tract. Upstream from the polypyrimidine tract is the branch point, which includes an adenine nucleotide. Point mutations in the underlying DNA or errors during transcription can activate a "cryptic splice site" in part of the transcript that usually is not spliced. This results in a mature messenger RNA with a missing section of an exon. In this way a point mutation, which usually only affects a single amino acid, can manifest as a deletion in the final protein.

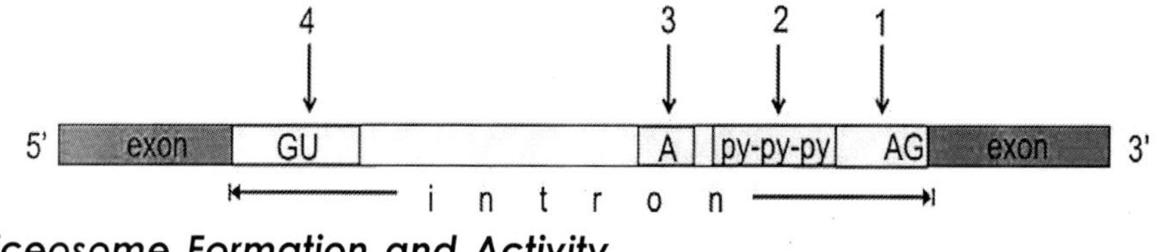

Spliceosome Formation and Activity

Splicing is catalyzed by the spliceosome which is a large RNA-protein complex composed of five small nuclear ribonucleoproteins (snRNPs, pronounced 'snurps').

The RNA components of snRNPs interact with the intron and may be involved in catalysis. Two types of spliceosomes have been identified (the major and minor) which contain different snRNPs.

Major

The major spliceosome splices introns containing GU at the 5' splice site and AG at the 3' splice site.

It is composed of the U1, U2, U4, U5, and U6 snRNPs and is active in the nucleus.

In addition, a number of proteins including U2AF and SF1 are required for the assembly of the spliceosome.

- E Complex-U1 binds to the GU sequence at the 5' splice site, along with accessory proteins/enzymes ASF/SF2, U2AF (binds at the Py-AG site), SF1/BBP (BBP=Branch Binding Protein);
- A Complex-U2 binds to the branch site and ATP is hydrolyzed;
- B1 Complex-U5/U4/U6 trimer binds, and the U5 binds exons at the 5' site, with U6 binding to U2;
- B2 Complex-U1 is released, U5 shifts from exon to intron and the U6 binds at the 5' splice site;
- C1 Complex-U4 is released, U6/U2 catalyzes transesterification, that make 5'end of introns ligate to the A on intron and form a lariat ,U5 binds exon at 3' splice site, and the 5' site is cleaved, resulting in the formation of the lariat;
- C2 Complex-U2/U5/U6 remain bound to the lariat, and the 3' site is cleaved and exons are ligated using ATP hydrolysis. The spliced RNA is released and the lariat debranches.

This type of splicing is termed canonical splicing or termed the lariat pathway, which accounts for more than 99% of splicing.

By contrast, when the intronic flanking sequences do not follow the GU-AG rule, noncanonical splicing is said to occur (see "minor spliceosome" below).

Minor

The minor spliceosome is very similar to the major spliceosome, however it splices out rare introns with different splice site sequences.

While the minor and major spliceosomes contain the same U5 snRNP, the minor spliceosome has different, but functionally analogous snRNPs for U1, U2, U4, and U6, which are respectively called U11, U12, U4atac, and U6atac. Like the major spliceosome, it is only found in the nucleus.

Trans-splicing

Trans-splicing is a form of splicing that joins two exons that are not within the same RNA transcript.

Self-splicing

Self-splicing occurs for rare introns that form a ribozyme, performing the functions of the spliceosome by RNA alone. There are three kinds of self-splicing introns, Group I, Group II and Group III. Group I and II introns perform splicing similar to the spliceosome without requiring any protein. This similarity suggests that Group I and II introns may be evolutionarily related to the spliceosome. Self-splicing may also be very ancient, and may have existed in an RNA world present before protein. Although the two splicing mechanisms described below do not require any proteins to occur, 5 additional RNA molecules and over 50 proteins are used and hydrolyzes many ATP molecules. The splicing mechanisms use ATP in order to accurately splice mRNA's. If the cell were to not use any ATP's, the process would be highly inaccurate and many mistakes would occur. Two transesterifications characterize the mechanism in which group I introns are spliced:

1. 3'OH of a free guanine nucleoside (or one located in the intron) or a nucleotide cofactor (GMP, GDP, GTP) attacks phosphate at the 5' splice site.
2. 3'OH of the 5'exon becomes a nucleophile and the second transesterification results in the joining of the two exons.

The mechanism in which group II introns are spliced (two transesterification reaction like group I introns) is as follows:

1. The 2'OH of a specific adenosine in the intron attacks the 5' splice site, thereby forming the lariat
2. The 3'OH of the 5' exon triggers the second transesterification at the 3' splice site thereby joining the exons together.

tRNA Splicing

tRNA (also tRNA-like) splicing is another rare form of splicing that usually occurs in tRNA. The splicing reaction involves a different biochemistry than the spliceomsomal and self-splicing pathways. Ribonucleases cleave the RNA and ligases join the exons together.

Evolution

Splicing occurs in all the kingdoms or domains of life, however, the extent and types of splicing can be very different between the major divisions. Eukaryotes splice many protein-coding messenger RNAs and some non-coding

RNAs. Prokaryotes, on the other hand, splice rarely and mostly non-coding RNAs. Another important difference between these two groups of organisms is that prokaryotes completely lack the spliceosomal pathway. Because spliceosomal introns are not conserved in all species, there is debate concerning when spliceosomal splicing evolved. Two models have been proposed: the intron late and intron early models (see intron evolution).

Splicing Diversity

	Eukaryotes	*Prokaryotes*
Spliceosomal	+	-
Self-splicing	+	+
tRNA	+	+

Biochemical Mechanism

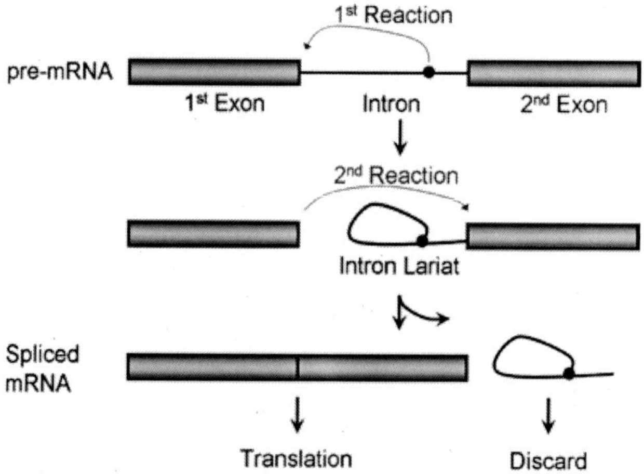

Figure: *Diagram illustrating the two-step biochemistry of splicing*

Spliceosomal splicing and self-splicing involves a two-step biochemical process. Both steps involve transesterification reactions that occur between RNA nucleotides. tRNA splicing, however, is an exception and does not occur by transesterification.

Spliceosomal and self-splicing transesterification reactions occur via two sequential transesterification reactions. First, the 2'OH of a specific branch-point nucleotide within the intron that is defined during spliceosome assembly performs a nucleophilic attack on the first nucleotide of the intron at the 5' splice site forming the lariat intermediate. Second, the 3'OH of the released 5' exon then performs a nucleophilic attack at the last nucleotide of the intron at the 3' splice site thus joining the exons and releasing the intron lariat.

Alternative Splicing

In many cases, the splicing process can create a range of unique proteins by varying the exon composition of the same messenger RNA. This phenomenon is then called alternative splicing. Alternative splicing can occur in many ways. Exons can be extended or skipped, or introns can be retained.

Experimental Manipulation of Splicing

Splicing events can be experimentally altered by binding steric-blocking antisense oligos such as Morpholinos or Peptide nucleic acids to snRNP binding sites, to the branchpoint nucleotide that closes the lariat, or to splice-regulatory element binding sites.

Splicing Errors

Common errors:

- Mutation of a splice site resulting in loss of function of that site. Results in exposure of a premature stop codon, loss of an exon, or inclusion of an intron.
- Mutation of a splice site reducing specificity. May result in variation in the splice location, causing insertion or deletion of amino acids, or most likely, a disruption of the reading frame.
- Displacement of a splice site, leading to inclusion or exclusion of more RNA than expected, resulting in longer or shorter exons.

Many splicing errors are safeguarded by a cellular quality control mechanism termed Nonsense-mediated mRNA decay [NMD].

Protein Splicing

In addition to RNA, proteins can undergo splicing. Although the biomolecular mechanisms are different, the principle is the same: parts of the protein, called inteins instead of introns, are removed. The remaining parts, called exteins instead of exons, are fused together. Protein splicing has been observed in a wide range of organisms, including bacteria, archaea, plants, yeast and humans.

Minor spliceosome

The minor spliceosome is a ribonucleoprotein complex that catalyses the removal (splicing) of an atypical class of spliceosomal introns (U12-type) from eukaryotic messenger RNAs in plant, insects, vertebrates and some fungi (Rhizopus oryzae). This process is called noncanonical splicing, as opposed to U2-dependent canonical splicing. U12-type introns represent less than 1% of all introns in human cells. However they are found in genes performing essential cellular functions.

Early Evidences

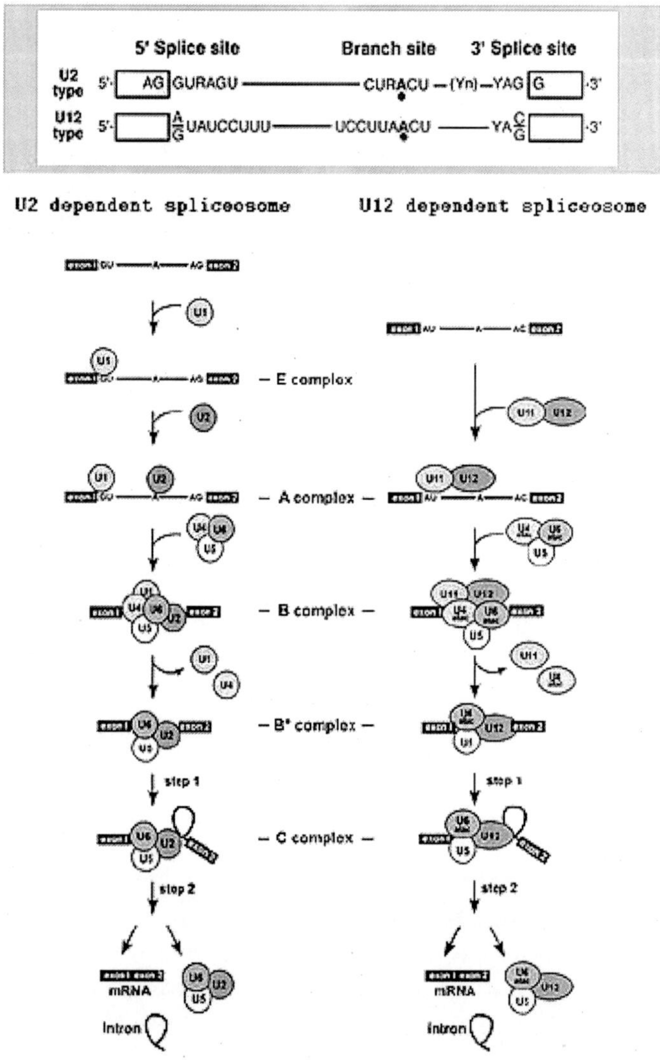

Figure: *A comparison between major and minor splicing mechanisms*

A notable feature of eukaryotic nuclear pre mRNA introns is the relatively high level of conservation of the primary sequences of 5' and 3' splice sites over a great range of organisms.

Since 1989 till 1991, several groups reported four independent examples of introns with a splice site that differed from the common intron:

- Cartilage matrix protein (CMP/MATN1) gene in human and chicken
- Proliferating cell nucleolar protein P120 (NOL1) gene in human

- mouse Rep3 gene, presumably involved in DNA repair
- Drosophila prospero gene that encodes for a homeobox protein

In 1991 by comparing the intron sequences of P120 and CMP genes, IJ Jackson reported the existence of ATATCC (5') and YYCAC (3') splice sites in these introns. The finding indicated a possible novel splicing mechanism.

In 1994, S.L. Hall and R.A Padgett compared the primary sequence of all reports on the four genes mentioned above. The results suggested a new type of introns with ATATCCTT 5' splice site and YCCAC 3' splice site and an almost invariant TCCTTAAC near the 3' end of the introns (so called 3' upstream element).

A search for small nuclear RNA sequences that are complementary to these splice sites, suggested U12 snRNA (matches 3' sequence) and U11 snRNA (matches 5'sequence) as being putative factors involved in splicing of this new type of introns.

In all these four genes, the pre-mRNA contains other introns whose sequences conform to those of major class introns. Neither the size nor the position of the AT–AC intron within the host gene is conserved.

In 1996, Woan-Yuh Tarn and Joan A. Steitz described an in vitro system that splices a pre-mRNA substrate containing an AT–AC intron derived from the human P120 gene.

Psoralen cross-linking confirms the base-pairing interaction predicted by Hall and Padgett between the branch site of the pre-mRNA substrate and U12 RNA.

Native gel electrophoresis reveals that U11, U12, and U5 snRNPs assemble onto the P120 pre-mRNA to form splicing complexes.

Structure of U12-type Introns and Minor Spliceosome

Although originally referred to as ATAC introns, U2-type introns have GT-AG 5' and 3' splice sites while U12-type introns have AT-AC at their 5' and 3' ends.

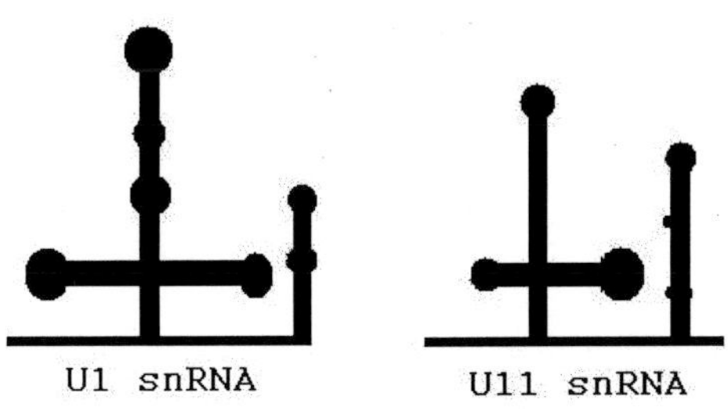

Figure: *U1 and U11 can be folded similarly*

The main determinants for distinguishing U2- and U12-type introns are 5' splice site and branch site sequences.

Minor spliceosome consists of U11, U12, U4atac, and U6atac, together with U5 and an unknown number of non-snRNP proteins. The U11, U12 and U4atac/U6atac snRNPs are functional analogs of the U1, U2 and U4/U6 snRNPs in major spliceosome.

Although the minor U4atac and U6atac snRNAs are functional analogs of U4 and U6, respectively, they share only limited sequence homology (ca. 40%). Furthermore, the sequence of U11 in comparison with U1, as well as U12 compared with U2, are completely unrelated. Despite this fact, the minor U11, U12, U4atac and U6atac snRNAs can be folded into structures similar to U1, U2, U4 and U6, respectively.

Location of Minor Spliceosomal Activity

The location of spliceosomal activity for the minor class spliceosome is regarded by most experts to be in the nucleus. However, a single paper has claimed that the minor spliceosome is active in the cytosol. The data presented within this paper are not fully accepted within the field and directly contradict numerous other papers.

Evolution of Minor Spliceosome

Like major spliceosome, minor spliceosome had an early origin: several of its characteristic constituents are present in representative organisms from all eukaryotic supergroups for which there is any substantial genome sequence information. In addition, functionally important sequence elements contained within U12-type introns and snRNAs are highly conserved during evolution.

Exon Junction Complex

The exon junction complex (EJC) has major influences on translation, surveillance and localization of the spliced mRNA. It is first deposited onto mRNA during splicing and is then transported into the cytoplasm. There it plays a major role in post-transcriptional regulation of mRNA.

It is believed that exon junction complexes provide a position-specific memory of the splicing event. The EJC consists of a stable heterotetramer core, which serves as a binding platform for other factors necessary for the mRNA pathway.

The core of the EJC contains the protein eukaryotic initiation factor 4AIII (eIF4AIII; a DEAD-box RNA helicase) bound to an adenosine triphosphate (ATP) analog, as well as the additional proteins Magoh and Y14. In order for the binding of the complex to the mRNA to occur, the eIF4AIII factor is inhibited, stopping the hydrolysis of ATP. This recognizes EJC as an ATP dependent complex.

Protein Components

The EJC is made up of several key protein components: RNPS1, Y14, SRm160, Aly/REF and Magoh, among others. RNPS1 can function as a coactivator of splicing, but along with Y14, it also takes part in the process of nonsense-mediated decay in eukaryotes. SRm160 is another coactivator that has been proposed to enhance mRNA 3' end processing. The protein component Magoh is thought to facilitate the subcytoplasmic localization of mRNAs while Aly is engaged in nuclear mRNA export. Aly is believed to be recruited to the exon junction complex by the protein UAP56. UAP56 is recognized as a RNA helicase but acts as a splicing factor required for early splicesome assembly. Another factor involved in the EJC pathway is DEK. This component is known to take part in a variety of functions ranging from splicing to transcriptional regulation and chromatin structure.

Structure

The crystallization of the exon junction complex has revealed the structural organization of its protein components. The core of the complex is elongated with an overall dimension of 99Å by 67Å by 54Å. It is organized around the eIF4AIII factor. The factor itself consists of two different types of conformations around the mRNA: closed and open. In a closed state, the two domains of eIF4AIII form composite binding sites for the 5'-adenylyl-ß- -imidodiphosphate (ADPNP) and mRNA.

In the open conformation, the two domains are rotated by 160 degrees relative to closed state18. The protein components Magoh and Y14 bind together to form a heterodimer located at the 5' pole of the EJC. Magoh binds to an eIF4AIII domain through interactions between residues from its two C-terminal helices and one end of a large â-sheet.

Conserved residues in the linker between the two eIF4AIII domains form salt bridges or hydrogen bonds with specific residues in Magoh. Other bonding occurs between the second loop of the Magoh â–sheet and the two eIF4AIII domains and their linker.

There is only a single partial bond formed between Y14 and eIF4AIII. This consists of a salt bridge between the conserved residues Y14 Arg108 and eIF4AIII Asp401.

If mutations were to occur to both of these residues, association of Magoh-Y14 with EJC would be non-existent.

Mechanism

During the second step of splicing in eukaryotic cells, the EJC is deposited approximately 20-24 nucleotides from the 5' end upstream of the splice junction (where two exons are joined), when the lariat has formed and the exons are ligated together.

The binding of the EJC to the mRNA occurs in a sequence independent manner, to form the mature messenger ribonucleoprotein (mRNP). The EJC remains stably bound to this mRNP as it is exported out of the nucleus and into the cytoplasm.

Protein components are either bound to or released by the EJC as it is transported. In order for the translocation of mRNAs through the nuclear pore complex to occur, a heterodimer consisting of NXF1/TAP and NXT1/p15 must bind to the transcripts. NXF1/TAP is a major receptor for the export of mRNAs to the cytoplasm. This is because it interacts with both RNA-binding adapter proteins and components of the nuclear pore complex.

EJC in Nonsense Mediated Decay

Exon junction complexes play a major role in mRNA surveillance. More specifically, they are found in the nonsense mediated decay pathway (NMD). In normal mRNA translation, the ribosome binds to the transcript and begins amino acid chain elongation. It continues on until it reaches the location of the exon junction complex, which it then displaces. Next, translation is complete when the ribosome reaches a termination codon. In NMD, the mRNA transcript contains a premature termination codon (PTC) due to a nonsense mutation. If this codon occurs prior to the EJC site, then mRNA decay is triggered. The EJC and its position serve as a type of regulator, determining whether the transcript is defective or not.

EJCs are also known to take part in NMD in another way; the recruitment of the surveillance factors UPF1, UPF2 and UPF3. These proteins are the most important components of the NMD mechanism. The EJC protein MAGOH, Y14 and eIF4AIII provide a binding for UPF3, which acts as a bridge between UPF2 and UPF1 forming a trimeric complex. Within this complex, UPF2 and UPF3 act cooperatively to promote ATPase and RNA helicase of UPF1. The EJC core stably anchors the UPF complex to the mRNA, and aids in regulation of essential UPF1 protein. Ribosomes which are stalled on a PTC recruit UPF1 through interactions with the release factor eRF1 and eRF3. Along with the protein SMG1, eRF1, eRF3 and UPF1 form the complex SURF. This complex forms a bridge between the ribosome and the downstream EJC which is associated with UPF3 and UPF2. This interaction triggers the phosphorylation of UPF1 by SMG1, causing the dissocation of eRF1 and eRF3. The complex produced consists of EJC, UPF3, UPF2, phosphorylated UPF1 and SMG1 and in turn triggers degradation of the mRNA.

Non-coding RNA Maturation

In most organisms non-coding genes (ncRNA) are transcribed as precursors which undergo further processing. In the case of ribosomal RNAs (rRNA), they are often transcribed as a pre-rRNA which contains one or more rRNAs, the pre-rRNA is cleaved and modified (22 -O-methylation and pseudouridine formation) at a specific sites by approximately 150 different small nucleolus-restricted RNA species, called snoRNAs.

SnoRNAs associate with proteins, forming snoRNPs. While snoRNA part basepair with the target RNA and thus position the modification to precise site, the protein part performs the catalytical reaction. In eukaryotes, in particular a snoRNP, called RNase MRP cleaves the 45S pre-rRNA into the 28S, 5.8S, and 18S rRNAs.

The rRNA and RNA processing factors form large aggregates called the nucleolus. In the case of transfer RNA (tRNA), for example, the 5' sequence is removed by RNase P, whereas the 3' end is removed by the tRNase Z enzyme and the non-templated 3' CCA tail is added by a nucleotidyl transferase. In the case of micro RNA (miRNA), miRNAs are first transcribed as primary transcripts or pri-miRNA with a cap and poly-A tail and processed to short, 70-nucleotide stem-loop structures known as pre-miRNA in the cell nucleus by the enzymes Drosha and Pasha.

After being exported, it is then processed to mature miRNAs in the cytoplasm by interaction with the endonuclease Dicer, which also initiates the formation of the RNA-induced silencing complex (RISC), composed of the Argonaute protein.

Even snRNAs and snoRNAs themselves undergo series of modification before they become part of functional RNP complex. This is done either in the nucleoplasm or in the specialized compartments called Cajal bodies. Their bases are methylated or pseudouridinilated by a group of small Cajal body-specific RNAs (scaRNAs) which are structurally similar to snoRNAs.

RNA Export

In eukaryotes most mature RNA must be exported to the cytoplasm from the nucleus. While some RNAs function in the nucleus, many RNAs are transported through the nuclear pores and into the cytosol. Notably this includes all RNA types involved in protein synthesis. In some cases RNAs are additionally transported to a specific part of the cytoplasm, such as a synapse; they are then towed by motor proteins that bind through linker proteins to specific sequences (called "zipcodes") on the RNA.

Translation

In molecular biology and genetics, translation is the third stage of protein biosynthesis (part of the overall process of gene expression). In translation, messenger RNA (mRNA) produced by transcription is decoded by the ribosome to produce a specific amino acid chain, or polypeptide, that will later fold into an active protein.

In Bacteria, translation occurs in the cell's cytoplasm, where the large and small subunits of the ribosome are located, and bind to the mRNA. In Eukaryotes, translation occurs across the membrane of the endoplasmic reticulum in a process called vectorial synthesis. The ribosome facilitates decoding by inducing the binding of tRNAs with complementary anticodon sequences to that of the mRNA. The tRNAs carry specific amino acids that are chained together into a polypeptide as the mRNA passes through and is "read" by the ribosome in a fashion reminiscent to that of a stock ticker and ticker tape.

In many instances, the entire ribosome/mRNA complex will bind to the outer membrane of the rough endoplasmic reticulum and release the nascent protein polypeptide inside for later vesicle transport and secretion outside of the cell. Many types of transcribed RNA, such as transfer RNA, ribosomal RNA, and small nuclear RNA, do not undergo translation into proteins.

Translation proceeds in four phases: activation, initiation, elongation and termination (all describing the growth of the amino acid chain, or polypeptide that is the product of translation). Amino acids are brought to ribosomes and assembled into proteins.

In activation, the correct amino acid is covalently bonded to the correct transfer RNA (tRNA). The amino acid is joined by its carboxyl group to the 3' OH of the tRNA by an ester bond. When the tRNA has an amino acid linked to it, it is termed "charged".

Initiation involves the small subunit of the ribosome binding to the 5' end of mRNA with the help of initiation factors (IF). Termination of the polypeptide happens when the A site of the ribosome faces a stop codon (UAA, UAG, or UGA). No tRNA can recognize or bind to this codon. Instead, the stop codon induces the binding of a release factor protein that prompts the disassembly of the entire ribosome/mRNA complex. A number of antibiotics act by inhibiting translation; these include anisomycin, cycloheximide, chloramphenicol, tetracycline, streptomycin, erythromycin, and puromycin, among others. Prokaryotic ribosomes have a different structure from that of eukaryotic ribosomes, and thus antibiotics can specifically target bacterial infections without any detriment to a eukaryotic host's cells.

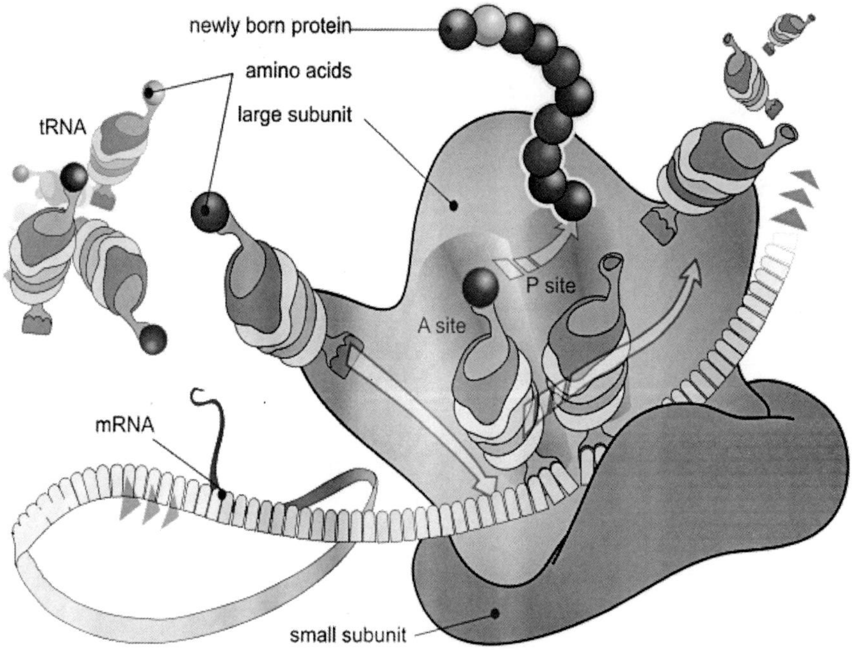

Figure: *During the translation, tRNA charged with amino acid enters the ribosome and aligns with the correct mRNA triplet. Ribosome then adds amino acid to growing protein chain.*

For some RNA (non-coding RNA) the mature RNA is the final gene product. In the case of messenger RNA (mRNA) the RNA is an information carrier coding for the synthesis of one or more proteins. mRNA carrying a single protein sequence (common in eukaryotes) is monocistronic whilst mRNA carrying multiple protein sequences (common in prokaryotes) is known as polycistronic.

Every mRNA consists of three parts - 5' untranslated region (5'UTR), protein-coding region or open reading frame (ORF) and 3' untranslated region (3'UTR). Coding region carries information for protein synthesis encoded by genetic code into form of triplets.

Each triplet of nucleotides of the coding region is called codon and corresponds to a binding site complementary to an anticodon triplet in transfer RNA. Transfer RNAs with the same anticodon sequence always carry identical type of amino acid.

Amino acids are then chained together by the ribosome according to order of triplets in the coding region. The ribosome helps transfer RNA to bind to messenger RNA and takes the amino acid from each transfer RNA and makes a structure-less protein out of it. Each mRNA molecule is translated into many protein molecules, on average ~900 in mammals.

In prokaryotes translation generally occurs at the point of transcription (co-transcriptionally), often using a messenger RNA which is still in the process of being created. In eukaryotes translation can occur in a variety of regions of the cell depending on where the protein being written is supposed to be.

Major locations are the cytoplasm for soluble cytoplasmic proteins and the membrane of endoplasmic reticulum for proteins which are for export from the cell or insertion into a cell membrane. Proteins which are supposed to be expressed at the endoplasmic reticulum are recognised part-way through the translation process. This is governed by the signal recognition particle - a protein which binds to the ribosome and directs it to the endoplasmic reticulum when it finds a signal sequence on the growing (nascent) amino acid chain. Basic mechanisms

The basic process of protein production is addition of one amino acid at a time to the end of a protein. This operation is performed by a ribosome. The choice of amino acid type to add is determined by an mRNA molecule. Each amino acid added is matched to a three nucleotide subsequence of the mRNA. For each such triplet possible, only one particular amino acid type is accepted. The successive amino acids added to the chain are matched to successive nucletide triplets in the mRNA. In this way the sequence of nucletides in the template mRNA chain determines the sequence of amino acids in the generated amino acid chain.

The mRNA carries genetic information encoded as a ribonucleotide sequence from the chromosomes to the ribosomes. The ribonucleotides are "read" by translational machinery in a sequence of nucleotide triplets called codons. Each of those triplets codes for a specific amino acid.

The ribosome molecules translate this code to a specific sequence of amino acids. The ribosome is a multisubunit structure containing rRNA and proteins. It is the "factory" where amino acids are assembled into proteins. tRNAs are small noncoding RNA chains (74-93 nucleotides) that transport amino acids to the ribosome. tRNAs have a site for amino acid attachment, and a site called an anticodon. The anticodon is an RNA triplet complementary to the mRNA triplet that codes for their cargo amino acid.

Aminoacyl tRNA synthetase (an enzyme) catalyzes the bonding between specific tRNAs and the amino acids that their anticodon sequences call for. The product of this reaction is an aminoacyl-tRNA molecule. This aminoacyl-tRNA travels inside the ribosome, where mRNA codons are matched through complementary base pairing to specific tRNA anticodons.

The ribosome has three sites for tRNA to bind. They are the aminoacyl site (abbreviated A), the peptidyl site (abbreviated P) and the exit site (abbreviated E). With respect to the mRNA, the three sites are oriented 5'to 3' E-P-A, because ribosomes move in a 3' to 5' fashion. The A site binds the incoming tRNA with the complementary codon on the mRNA. The P site holds the tRNA with the growing polypeptide chain. The E site holds the tRNA without its amino acid. When an aminoacyl-tRNA initially binds to its corresponding codon on the mRNA, it is in the A site. Then, a peptide bond forms between the amino acid of the tRNA in the A site and the amino acid of the charged tRNA in the P site. The growing polypeptide chain is transferred to the tRNA in the A site. Translocation occurs, moving the tRNA in the P site, now without an amino acid, to the E site; the tRNA that was in the A site, now charged with the polypeptide chain, to the P site. The tRNA in the E site leaves and another aminoacyl-tRNA enters the A site to repeat the process.

After the new amino acid is added to the chain, the energy provided by the hydrolysis of a GTP bound to the translocase EF-G (in prokaryotes) and eEF-2 (in eukaryotes) moves the ribosome down one codon towards the 3' end. The energy required for translation of proteins is significant. For a protein containing n amino acids, the number of high-energy Phosphate bonds required to translate it is 4n-1. The rate of translation varies; it is significantly higher in prokaryotic cells (up to 17-21 amino acid residues per second) than in eukaryotic cells (up to 6-9 amino acid residues per second).

5

Embryo Cryopreservation

Cryopreservation of embryos is the process of preserving an embryo at sub-zero temperatures, generally at an embryogenesis stage corresponding to pre-implantation, that is, from fertilization to the blastocyst stage.

Embryo cryopreservation is useful for leftover embryos after a cycle of in vitro fertilization, as patients who fail to conceive may become pregnant using such embryos without having to go through a full IVF cycle. Or, if pregnancy occurred, they could return later for another pregnancy. Spare oocytes or embryos resulting from fertility treatments may be used for oocyte donation or embryo donation to another woman or couple, and embryos may be created, frozen and stored specifically for transfer and donation by using donor eggs and sperm.

World usage data is hard to come by but it was reported in a study of 23 countries that almost 42,000 frozen human embryo transfers were performed during 2001 in Europe.

Pregnancy Outcome and Determinants

The outcome from using cryopreserved embryos has uniformly been positive with no increase in birth defects or development abnormalities, also between fresh versus frozen eggs used for intracytoplasmic sperm injection (ICSI). Children born from vitrified blastocysts have significantly higher birthweight than those born from non-frozen blastocysts. For early cleavage embryos, frozen ones appear to have at least as good obstetric outcome, measured as preterm birth and low birthweight for children born after cryopreservation as compared with children born after fresh cycles.

Oocyte age, survival proportion, and number of transferred embryos are predictors of pregnancy outcome.

Pregnancies have been reported from embryos stored for 16 years. A study of more than 11,000 cryopreserved human embryos showed no significant effect of storage time on post-thaw survival for IVF or oocyte donation cycles, or for embryos frozen at the pronuclear or cleavage stages. In addition, the duration of storage had no significant effect on clinical pregnancy, miscarriage, implantation, or live birth rate, whether from IVF or oocyte donation cycles.

Legislation

From October 1, 2009 human embryos are allowed to be stored for 10 years in the UK, according to the Human Fertilization and Embryology Act 2008.

History

The first ever pregnancy derived from a frozen human embryo was reported by Alan Trounson and Linda Mohr in 1983 (although the fetus aborted spontaneously at about 20 weeks of gestation); the first term pregnancies derived from frozen human frozen freezing process was born in 1984. Since then and up to 2008 it is estimated that between 350,000 and half a million IVF babies have been born from embryos frozen at a controlled rate and then stored in liquid nitrogen; additionally a few hundred births have been born from vitrified oocytes but firm figures are hard to come by.

It may be noted that Subash Mukhopadyay from Kolkata, India reported the successful cryopreservation of an eight cell embryo, storing it for 53 days, thawing and replacing it into the mother's womb, resulting in a successful and live birth as early as 1978- a full five years before Trounson and Mohr had done so. A small publication of Mukherjee in 1978 clearly shows that Mukherjee was on the right line of thinking much before anyone else had demonstrated the successful outcome of a pregnancy following the transfer of a 8-cell frozen-thawed embryo into human subjects transferring 8-cell cryopreserved embryos."

Other Expansions

- Intracytoplasmic sperm injection (*ICSI*) is where a single sperm is injected directly into an egg. Its main usage as an expansion of IVF is to overcome male infertility problems, although it may also be used where eggs cannot easily be penetrated by sperm, and occasionally in conjunction with sperm donation. It can be used in teratozoospermia, since once the egg is fertilized abnormal sperm morphology does not appear to influence blastocyst development or blastocyst morphology.

- Assisted zona hatching (AZH) can be performed shortly before the embryo is transferred to the uterus. A small opening is made in the outer layer surrounding the egg in order to help the embryo hatch out and aid in the implantation process of the growing embryo.
- In egg donation and embryo donation, the resultant embryo after fertilization is inserted in another woman than the one providing the eggs. These are resources for women with no eggs due to surgery, chemotherapy, or genetic causes; or with poor egg quality, previously unsuccessful IVF cycles or advanced maternal age. In the egg donor process, eggs are retrieved from a donor's ovaries, fertilized in the laboratory with the sperm from the recipient's partner, and the resulting healthy embryos are returned to the recipient's uterus.
- Embryo splitting can be used for twinning to increase the number of available embryos.

Leftover Embryos or Eggs

There may be leftover embryos or eggs from IVF procedures if the woman for whom they were originally created has successfully carried one or more pregnancies to term. With the woman's or couple's permission, these may be donated to help other women or couples as a means of third party reproduction. In embryo donation, these extra embryos are given to other couples or women for transfer with the goal of producing a successful pregnancy. The resulting child is considered the child of the woman who carries it and gives birth, and not the child of the donor, the same as occurs with egg donation or sperm donation. Typically, genetic parents donate the eggs to a fertility clinic or embryo bank where they are preserved by oocyte cryopreservation or embryo cryopreservation until a carrier is found for them. Typically the process of matching the embryo(s) with the prospective parents is conducted by the agency itself, at which time the clinic transfers ownership of the embryos to the prospective parents.

In the United States, women seeking to be an embryo recipient undergo infectious disease screening required by the U.S. Food and Drug Administration (FDA), and reproductive tests to determine the best placement location and cycle timing before the actual Embryo Transfer occurs. The amount of screening the embryo has already undergone is largely dependent on the genetic parents' own IVF clinic and process. The embryo recipient may elect to have her own embryologist conduct further testing.

Alternatives to donating unused embryos are destroying them (or having them implanted at a time where pregnancy is very unlikely), keeping them frozen indefinitely, or donating them for use in research (with results in their unviability). Individual moral views on disposing leftover embryos may depend on personal views on the beginning of human personhood and definition and/ or value of potential future persons and on the value that is given to fundamental research questions. Some people believe donation of leftover embryos for research is a good alternative to discarding the embryos when patients receive proper, honest and clear information about the research project, the procedures and the scientific values).

History

There was a transient biochemical pregnancy reported by Australian Foxton School researchers in 1953. John Rock was the first to extract an intact fertilized egg. The first pregnancy achieved through *in vitro* human fertilization of a human oocyte was reported in The Lancet from the Monash University team in 1973, although it lasted only a few days and would today be called a biochemical pregnancy. There was also an ectopic pregnancy reported by Patrick Steptoe and Robert Edwards in 1976.

In 1977, Steptoe and Edwards successfully carried out a pioneering conception which resulted in the birth of the world's first baby to be conceived by IVF, Louise Brown on 25 July 1978, in Oldham General Hospital, Greater Manchester, UK. In October 1978, it was reported that Subash Mukhopadyay, a relatively unknown physician from Kolkata, India was performing experiments on his own with primitive instruments and a household refrigerator and this resulted in a test tube baby, later named as "Durga" (alias Kanupriya Agarwal) who was born on October 3, 1978. However, state authorities prevented him from presenting his work at scientific conferences and, in the absence of scientific evidence, his work is not recognised by the international scientific community.

Steptoe and Edwards were responsible for the world's second (confirmed) baby conceived by IVF, Alastair MacDonald born on 14 January 1979 in Glasgow. A team led by Ian Johnston and Alex Lopata were responsible for Australia's first baby conceived by IVF, Candice Reed born on 23 June 1980 in Melbourne. It was the subsequent use of stimulated cycles with clomiphene citrate and the use of human chorionic gonadotrophin (hCG) to control and time oocyte maturation, thus controlling the time of collection, that converted IVF from a research tool to a clinical treatment. This was followed by a total of 14 pregnancies resulting in nine births in 1981 with the Monash University team. The Jones team at the Eastern Virginia Medical School in Norfolk, Virginia, further improved stimulated cycles by incorporating the use of a follicle-stimulating hormone (uHMG). This then became known as controlled ovarian hyperstimulation (COH). Another step forward was the use of gonadotrophin-releasing hormone agonists (GnRHA), thus decreasing the need for monitoring by preventing premature ovulation, and more recently gonadotrophin-releasing hormone antagonists (GnRH Ant), which have a similar function. The additional use of the oral contraceptive pill has allowed the scheduling of IVF cycles, which has made the treatment far more convenient for both staff and patients.

The ability to freeze and subsequently thaw and transfer embryos has significantly improved the feasibility of IVF use. The other very significant milestone in IVF was the development of the intracytoplasmic sperm injection (ICSI) of single sperms by André van Steirteghem in Brussels, 1992. This has enabled men with minimal sperm production to achieve pregnancies. ICSI is sometimes used in conjunction with sperm recovery, using a testicular fine needle or open testicular biopsy.

Using this method, some men with Klinefelter's syndrome, and so would be otherwise infertile, have occasionally been able to achieve pregnancy. Thus, IVF has become the final solution for most fertility problems, moving from tubal disease to male factor, idiopathic subfertility, endometriosis, advanced maternal age, and anovulation not responding to ovulation induction. Robert Edwards was awarded the 2010 Nobel Prize in Physiology or Medicine "for the development of *in vitro* fertilization". Carl Wood was dubbed "the father of IVF (*in vitro* fertilization)" for having pioneered the use of frozen embryos.

In the US, ART cycles started in 2006 resulted in 41,343 births (54,656 infants), which is slightly more than 1% of total US births.

Ethics

Issues

In a few cases, laboratory mix-ups (misidentified gametes, transfer of wrong embryos) have occurred, leading to legal action against the IVF provider and complex paternity suits. An example is the case of a woman in California who received the embryo of another couple and was notified of this mistake after the birth of her son. This has led to many authorities and individual clinics implementing procedures to minimise the risk of such mix-ups. The HFEA, for example, requires clinics to use a double witnessing system, where the identity of specimens is checked by two people at each point at which specimens are transferred. Alternatively, technological solutions are gaining favour, to reduce the manpower cost of manual double witnessing, and to further reduce the risk of human error. Technological solutions typically involve tagging individual specimen containers with uniquely numbered RFID tags which can be identified by readers connected to a computer. The computer tracks specimens throughout the process and alerts the embryologist if non-matching specimens are identified.

Another concern is that people will screen in or out for particular traits, using preimplantation genetic diagnosis. For example, a deaf British couple, Tom and Paula Lichy, have petitioned to create a deaf baby using IVF. Some medical ethicists have been very critical of this approach. Jacob Appel wrote that "intentionally culling out blind or deaf embryos might prevent considerable future suffering, while a policy that allowed deaf or blind parents to select *for* such traits intentionally would be far more troublesome."

Pregnancy: Past-Menopause

Although menopause is a natural barrier to further conception, IVF has allowed women to be pregnant in their fifties and sixties. Women whose uterus has been appropriately prepared receive embryos that originated from an egg of an egg donor. Therefore, although these women do not have a genetic link with the child, they have an emotional link through pregnancy and childbirth. In many cases the genetic father of the child is the woman's partner. Even after menopause the uterus is fully capable of carrying out a pregnancy.

Same-sex Couples, Single and Unmarried Parents

A 2009 statement from the ASRM found no persuasive evidence that children are harmed or disadvantaged solely by being raised by single parents, unmarried parents, or homosexual parents. It did not support restricting access to assisted reproductive technologies on the basis of a prospective parent's marital status or sexual orientation.

Ethical concerns include reproductive rights, the welfare of offspring, nondiscrimination against unmarried individuals, homosexual, and professional autonomy.

A recent controversy in California focused on the question of whether physicians opposed to same-sex relationships should be required to perform IVF for a lesbian couple. Guadalupe T. Benitez, a medical assistant from San Diego, sued doctors Christine Brody and Douglas Fenton of the North Coast Women's Care Medical Group after Brody told her that she hd "religious-based objections to treating homosexuals to help them conceive children by artificial insemination," and Fenton refused to authorise a refill of her prescription for the fertility drug Clomid on the same grounds. The California Medical Association had initially sided with Brody and Fenton, but the case, North Coast Women's Care Medical Group vs Superior Court, was decided unanimously by the California State Supreme Court in favour of Benitez on August 19, 2008.

Nadya Suleman came to international attention after having twelve embryos implanted, eight of which survived, resulting in eight newborns being added to her existing six-child family. The Medical Board of California sought to have fertility doctor Michael Kamrava, who treated Suleman, stripped of his license. State officials allege that performing Suleman's procedure is evidence of unreasonable judgement, substandard care, and a lack of concern for the eight children she would conceive and the six she was already struggling to raise. On June 1, 2011 the Medical Board issued a ruling that Kamrava's medical licence be revoked effective July 1, 2011.

Catholic Objections

The Catholic Church opposes all kinds of in vitro fertilization because, as with contraception, it separates the procreative purpose of the marriage act from its unitive purpose: This particular doctrine, often expounded by the magisterium of the Church, is based on the inseparable connection, established by God, which man on his own initiative may not break, between the unitive significance and the procreative significance which are both inherent to the marriage act.

The reason is that the fundamental nature of the marriage act, while uniting husband and wife in the closest intimacy, also renders them capable of generating new life—and this as a result of laws written into the actual nature of man and of woman. And if each of these essential qualities, the unitive and the procreative, is preserved, the use of marriage fully retains its sense of true mutual love and its ordination to the supreme responsibility of parenthood to which man is called. We believe that our contemporaries are particularly capable of seeing that this teaching is in harmony with human reason.

According to the Catechism of the Catholic Church, techniques involving only the married couple (homologous artificial insemination and fertilization) are perhaps less reprehensible, yet remain morally unacceptable. They dissociate the sexual act from the procreative act. The act which brings the child into existence is no longer an act by which two persons give themselves to one another, but one that "entrusts the life and identity of the embryo into the power of doctors and biologists and establishes the domination of technology over the origin and destiny of the human person. Such a relationship of domination is in itself contrary to the dignity and equality that must be common to parents and children."

The Catholic Church maintains that it is not objectively evil to be infertile, and advocates adoption as an option for such couples who still wish to have children: The Gospel shows that physical sterility is not an absolute evil. Spouses who still suffer from infertility after exhausting legitimate medical procedures should unite themselves with the Lord's Cross, the source of all spiritual fecundity. They can give expression to their generosity by adopting abandoned children or performing demanding services for others.

Gamete Intrafallopian Transfer (GIFT) is not technically *in vitro* fertilization because with GIFT, fertilization takes place inside the body, not on a Petri dish. The Catholic Church nevertheless is concerned with it because "Some theologians consider this to be a replacement of the marital act, and therefore immoral."

Availability and Utilization

In the USA, overall availability of IVF in 2005 was 2.5 IVF physicians per 100,000 population, and utilization was 236 IVF cycles per 100,000. Utilisation highly increases with availability and IVF insurance coverage, and to a significant extent also with percentage of single persons and median income. The cost of IVF rather reflects the costliness of the underlying health care system than the regulatory or funding environment, and ranges, on average for a standard IVF cycle and in 2006 United States dollars, between $12,500 in the United States to $4,000 in Japan. In Ireland, IVF costs around €4,000, with fertility drugs, if required, costing up to €3,000. The cost per live birth is highest in the United States ($41,000) and United Kingdom ($40,000) and lowest in Scandinavia and Japan (both around $24,500).

Many fertility clinics in the United States limit the upper age at which women are eligible for IVF to 50 or 55 years. These cutoffs make it difficult for women older than fifty-five to utilise the procedure.

In Australia, the average age of women undergoing ART treatment is 35.5 years among those using their own eggs (one in four being 40 or older) and 40.5 years among those using donated eggs.

Legal Status

Government agencies in China passed bans on the use of IVF in 2003 by unmarried women or by couples with certain infectious diseases. Sunni Muslim nations generally allow IVF between married couples when conducted with their own respective sperm and eggs, but not with donor eggs from other couples. But Iran, which is Shi'a Muslim, has a more complex scheme. Iran bars sperm donation but allows donation of both fertilized and unfertilized eggs. Fertilized eggs are donated from married couples to other married couples, while unfertilized eggs are donated in the context of mut'ah or temporary marriage to the father. The nation of Costa Rica has a complete ban on all

IVF technology, it having been ruled unconstitutional by the nation's Supreme Court because it "violated life." Costa Rica is the only country in the western hemisphere that forbids this technique.

The Inter-American Court on Human Rights ruled out that the Costa Rican government has violated the human rights of infertile couples when it banned the use of this technique.

According to the rule, Costa Rica had to enact legislation allowing the use of this type of procedure. The Court set a first dateline on February 2011 for a report on the question by the Costa Rican government.

The latter missed the dateline and new datelines were set. All of them have been missed by the Costa Rican government. A law project sent reluctantly by the government of Pres. Laura Chinchilla was rejected at the Costa Rican parliament. Consequently, observers expect that Costa Rica will not abide to the request of the Inter-American Court on Human Rights.

President Chinchilla, whose strong Catholic views have won her to be named officially as Preferred Daughter of the Virgin Mary has not publicly stated her position on the question of *in vitro* fertilization.

Federal regulations in the United States include screening requirements and restrictions on donations, but generally do not affect sexually intimate partners. However, doctors may be required to *provide* treatments due to nondiscrimination laws, as for example in California.

All major restrictions on single but infertile women using IVF were lifted in Australia in 2002 after a final appeal to the Australian High Court was rejected on procedural grounds in the Leesa Meldrum case. A Victorian federal court had ruled in 2000 that the existing ban on all single women and lesbians using IVF constituted sex discrimination.

Victoria's government announced changes to its IVF law in 2007 eliminating remaining restrictions on fertile single women and lesbians, leaving South Australia as the only state maintaining them.

The US state of Tennessee proposed a bill in 2009 that would have defined donor IVF as adoption. During the same session another bill proposed barring adoption from any unmarried and cohabitating couple, and activist groups stated that passing the first bill would effectively stop unmarried people from using IVF. Neither of these bills passed.

Assisted Reproduction

Assisted reproductive technology (ART) is a general term referring to methods used to achieve pregnancy by artificial or partially artificial means.

It is reproductive technology used primarily in infertility treatments. Some forms of *ART* are also used in fertile couples for genetic reasons. *ART* is also used in couples who are discordant for certain communicable diseases, e.g. AIDS, to reduce the risk of infection when a pregnancy is desired. The term includes any reproductive technique involving a third party e.g. a sperm donor. Examples of *ART* include in vitro fertilization, intracytoplasmic sperm injection (ICSI), cryopreservation, and intrauterine insemination (IUI). There is yet no strict definition of the term. Usage of the ART mainly belongs in the field of reproductive endocrinology and infertility.

Definitions

While there is no consensus on the definition, generally the process of intercourse is bypassed either by insemination (for example, artificial insemination) or fertilization of the oocytes in the laboratory environment (i.e., in vitro fertilization).

- The Centres for Disease Control and Prevention (CDC)—which is required as a result of the 1992 Fertility Clinic Success Rate and Certification Act to publish the annual ART success rates at U.S. fertility clinics—defines ART to include "all fertility treatments in which both eggs and sperm are handled. In general, ART procedures involve surgically removing eggs from a woman's ovaries, combining them with sperm in the laboratory, and returning them to the woman's body or donating them to another woman." According to CDC, "they do not include treatments in which only sperm are handled (i.e., intrauterine—or artificial—insemination) or procedures in which a woman takes medicine only to stimulate egg production without the intention of having eggs retrieved."

Procedures

Procedures are mainly fertility medication, as well as ART techniques that use more substantial and forceful interventions, of which in vitro fertilization (IVF) and expansions of it (e.g. OCR, AZH, ICSI, ZIFT) are the most prevalent. However, there are also other manual ART, not necessarily dependent on IVF (e.g. PGD, GIFT, SSR).

Medication

Most fertility medication are agents that stimulate the development of follicles in the ovary. Examples are gonadotropins and gonadotropin releasing hormone.

In vitro Fertilization

In vitro fertilization (IVF) is the technique of letting fertilization of the male and female gametes (sperm and egg) occur outside the female body.

Techniques usually used in vitro fertilization include:

- Transvaginal ovum retrieval (OCR) is the process whereby a small needle is inserted through the back of the vagina and guided via ultrasound into the ovarian follicles to collect the fluid that contains the eggs.
- Embryo transfer is the step in the process whereby one or several embryos are placed into the uterus of the female with the intent to establish a pregnancy.

Expansions of IVF

The following are techniques generally requires methods of in vitro fertilization. In vitro fertilization, however, usually does not require these techniques.

- Assisted zona hatching (AZH) is performed shortly before the embryo is transferred to the uterus. A small opening is made in the outer layer surrounding the egg in order to help the embryo hatch out and aid in the implantation process of the growing embryo.
- Intracytoplasmic sperm injection (ICSI) is beneficial in the case of male factor infertility where sperm counts are very low or failed fertilization occurred with previous IVF attempt(s). The ICSI procedure involves a single sperm carefully injected into the centre of an egg using a microneedle. This method is also sometimes employed when donor sperm is used.
- Autologous endometrial coculture is a possible treatment for patients who have failed previous IVF attempts or who have poor embryo quality. The patient's fertilized eggs are placed on top of a layer of cells from the patient's own uterine lining, creating a more natural environment for embryo development.
- In zygote intrafallopian transfer (ZIFT), egg cells are removed from the woman's ovaries and fertilized in the laboratory; the resulting zygote is then placed into the fallopian tube.
- Cytoplasmic transfer is the technique in which the contents of a fertile egg from a donor are injected into the infertile egg of the patient along with the sperm.
- Egg donors are resources for women with no eggs due to surgery, chemotherapy, or genetic causes; or with poor egg quality, previously unsuccessful IVF cycles or advanced maternal age. In the egg donor process, eggs are retrieved from a donor's ovaries, fertilized in the

laboratory with the sperm from the recipient's partner, and the resulting healthy embryos are returned to the recipient's uterus.

- Sperm donation may provide the source for the sperm used in IVF procedures where the male partner produces no sperm or has an inheritable disease, or where the woman being treated has no male partner.
- A gestational carrier is an option when a patient's medical condition prevents a safe pregnancy, when a patient has ovaries but no uterus due to congenital absence or previous surgical removal, and where a patient has no ovaries and is also unable to carry a pregnancy to full term.
- Preimplantation genetic diagnosis (PGD) involves the use of genetic screening mechanisms such as Fluorescent In Situ Hybridisation (FISH) or Comparative Genomic Hybridisation (CGH) to help identify genetically abnormal embryos and improve healthy outcomes.
- Embryo splitting can be used for twinning to increase the number of available embryos.

Others

The following Assisted Reproduction techniques don't necessarily involve IVF.

- In gamete intrafallopian transfer (GIFT) a mixture of sperm and eggs is placed directly into a woman's fallopian tubes using laparoscopy following a transvaginal ovum retrieval.
- Sex selection is the attempt to control the sex of offspring to achieve a desired sex. It can be accomplished in several ways, both pre- and post-implantation of an embryo, as well as at birth. Pre-implantation techniques include PGD, but also sperm sorting.
- Artificial insemination (AI) is when sperm is placed into a female's uterus (intrauterine) or cervix (intracervical) using artificial means rather than by natural copulation. N.B. This can be a very low-tech process, performed at home by the woman alone or with her partner.
    - o Conception devices, such as a conception cap are used to aid conception by enhancing the natural process. Conception caps are used by placing semen into a small conception cap, then placing the cap onto the cervix. This holds the semen at the cervical os, protecting the semen from the acidic vaginal secretions and keeping it in contact with the cervical mucus.

  - o Artificial insemination by donor is used in situations where the woman doesn't have a partner with functional sperm. Instead, a sperm donor supplies the sperm.
- Surrogacy, where a woman agrees to become pregnant and deliver a child for a contracted party. It may be her own genetic child, or a child conceived through in vitro fertilization or embryo transfer using another woman's ova.
- Reproductive surgery, treating e.g. fallopian tube obstruction and vas deferens obstruction, or reversing a vasectomy by a reverse vasectomy.
  - o In surgical sperm retrieval (SSR) the reproductive urologist obtains sperm from the vas deferens, epididymis or directly from the testis in a short outpatient procedure.
- By cryopreservation, eggs, sperm and reproductive tissue can be preserved for later IVF.

Risks

The majority of IVF-conceived infants do not have birth defects. However, some studies have suggested that assisted reproductive technology is associated with an increased risk of birth defects. In the largest U.S. study, which used data from a statewide registry of birth defects, 6.2% of IVF-conceived children had major defects, as compared with 4.4% of naturally conceived children matched for maternal age and other factors (odds ratio, 1.3; 95% confidence interval, 1.00 to 1.67).

The main risks are:

- Genetic disorders. DNA damage increases in e.g. IVF and ICSI, which is reflected e.g. by upregulation of the gene expression of HNRNPC in the placenta.
- Low birth weight. In IVF and ICSI, a risk factor is the decreased expression of proteins in energy metabolism; Ferritin light chain and ATP5A1.
- Preterm birth. Low birth weight and preterm birth are strongly associated with many health problems, such as visual impairment and cerebral palsy, and children born after IVF are roughly twice as likely to have cerebral palsy.

Other risk factors are:

- Membrane damage, which is contributed to or reflected by increased expression of the membrane fusion proteins NAPA and Annexin A3.

Sperm donation is an exception, with a birth defect rate of almost a fifth compared to the general population. It may be explained by that sperm banks accept only people with high sperm count.

Current data indicate little or no increased risk for postpartum depression among women who use ART.

Usage

Assisted reproductive technology procedures performed in the U.S. has more than doubled since 10 years ago, with 140.000 procedures in 2006, resulting in 55.000 infants born. In Australia, 3.1 percent of babies now born are a result of ART.

Costs

United States of America

Not everyone in the U.S. has insurance coverage for fertility investigations and treatments. Many states are starting to mandate coverage, and the rate of utilisation is 277% higher in states with complete coverage. There are some health insurance companies that cover diagnosis of infertility but frequently once diagnosed will not cover any treatment costs. 2005 approximate treatment/ diagnosis costs (United States, costs in US$):

- Initial workup: hysteroscopy, hysterosalpingogram, blood tests ~$2,000
- Intrauterine Insemination (IUI) aka Artificial insemination ~ $200–900 per. trial
- Sonohysterogram (SHG) ~ $600 – 1,000
- Clomiphene citrate cycle ~ $ 200 - 500
- IVF cycle ~ $10,000 -30,000
- Use of a surrogate mother to carry the child - dependent on arrangements.

Another way to look at costs is to determine the expected cost of establishing a pregnancy. Thus if a clomiphene treatment has a chance to establish a pregnancy in 8% of cycles and costs $500, the expected cost is $6,000 to establish a pregnancy, compared to an IVF cycle (cycle fecundity 40%) with a corresponding expected cost of $30,000 ($12,000/.4). For the community as a whole, the cost of IVF on average pays back by 700% by tax from future employment by the conceived human being.

United Kingdom

In the UK all patients have the right to preliminary testing, provided free of charge by the National Health Service. However, treatment is not widely available on the NHS and there can be long waiting lists. Many patients therefore pay for immediate treatment within the NHS or seek help from private clinics.

Sweden

In Sweden, official fertility clinics provide most necessary treatments and initial workup, but there are long waiting lists, especially for egg donations, since the donor gets just as low reward as the receiving couple are charged. However, there are private fertility clinics.

Canada

Some treatments are covered by OHIP (public health insurance) in Ontario and others are not. Those undergoing artificial insemination or with bilaterally blocked fallopian tubes and under 40 have treatment is covered but are still required to pay lab fees (around $3,000-4,000). Coverage veries in other provinces. Most other patients are required to pay for treatments themselves.

Israel

Israel's National Health Insurance, which is mandatory for all Israeli citizens, covers nearly all fertility treatments. In-vitro-fertilization costs are fully subsidized up to the birth of two children for all Israeli women, including single women and lesbian couples. Embryo transfers for purposes of gestational surrogacy are also covered.

New Zealand

The national public health system of New Zealand covers IVF treatment in specific circumstances only, based on a 'points for conception challenges' equation. Publicly funded IVF treatments are limited (between one and three treatments dependent on criteria) and are subject to substantial wait-lists, dependent on local health funding region, which raises potential inequity of ART support across the country.

Infertility testing through blood tests can be covered by public funding, however in the absence of explicit gynecological complications, additional investigations are may not be covered publicly. Investigation such as a hysterosalpingogram may be covered, but the wait-list could be in excess of six weeks, whereas a privately sourced HSG can cost $NZ900 but is readily available.

Many New Zealanders select self-funded IVF cycles, at approximately $NZ10,000 per cycle, and other forms of ART, such as IUI, at approximately $NZ1200, using the services of private fertility clinics, which in itself is a growing local industry. Individuals using private services are generally not covered under personal health insurance policies in New Zealand.

Ethics

Some couples find it difficult to stop treatment despite very bad prognosis, resulting in futile therapies. This may give ART providers a difficult decision of whether to continue or refuse treatment.

Fictional Representation

Films and other fiction depicting emotional struggles of assisted reproductive technology have had an upswing in the latter part of the 2000s decade, although the techniques have been available for decades. Yet, the amount of people that can relate to it by personal experience in one way or another is ever growing, and the variety of trials and struggles are huge. For specific examples, refer to the fiction sections in individual subarticles, e.g. surrogacy, sperm donation and fertility clinic.

In addition, reproduction and pregnancy in speculative fiction has been present for many decades.

Sperm Bank

A sperm bank, semen bank or cryobank is a facility that collects and stores human sperm mainly from sperm donors, primarily for the purpose of achieving pregnancies through third party reproduction, notably by artificial insemination. Sperm donated in this way is known as donor sperm. The development of sperm banks, whilst considered by some to be controversial, has enabled people to have greater control over their reproductive lives. By providing sperm from donors who are checked and screened, thousands of women every year are able to bear their own children in circumstances where this might otherwise not be possible. Controversy stems from the fact that donors effectively father children for others, often in considerable numbers, and usually take no part in the upbringing of such children, and also from the fact that single women and coupled lesbians frequently use sperm banks in order to have their own biological children.

The increasing range of services which is available through sperm banks nevertheless enables more and more people to have choices over the whole issue of reproduction. Women may choose to use an anonymous donor who will not be a part of family life, or they may choose known donors who may be contacted later in life by the donor children. Women may choose to use a surrogate to bear their children, using eggs provided by the woman and sperm from a donor. Sperm banks often provide services which enable a woman to have subsequent pregnancies by the same donor, but equally, women may choose to have children by a number of different donors. Sperm banks sometimes enable a woman to choose the sex of her child, enabling even greater control over the way families are planned. Sperm banks increasingly adopt a less formal approach to the provision of their services thereby enabling people to take a relaxed approach to their own individual requirements. Men who choose to donate sperm through a sperm bank also have the security of knowing that they are helping women or childless couples to have children in circumstances where they, as the biological father, will not have any legal or other responsibility for the children produced from their sperm.

Whether a donor is anonymous or not, this factor is important in allowing sperm banks to recruit sperm donors and to use their sperm to produce whatever number of pregnancies from each donor as are permitted where they operate, or alternatively, whatever number they decide.

However, in many parts of the world sperm banks are not allowed to be established or to operate. Sperm banks do not provide a cure for infertility in that it is the sperm donor who reproduces himself, not a partner of the recipient woman. Most societies are built upon the family model and sperm banks may be seen as a threat to this, particularly where a sperm bank makes its services available to unmarried women.

Where sperm banks are allowed to operate they are often controlled by local legislation which is primarily intended to protect the unborn child, but which may also provide a compromise between the conflicting views which surround their operation.

A particular example of this is the control which is often placed on the number of children which a single donor may father and which may be designed to protect against consanguity: low limits are often provided which in reality appear more intended to placate those who would be opposed to sperm donors producing tens or scores of children as to provide a justifiable safeguard.

However, such legislation usually cannot prevent a sperm bank from supplying donor sperm outside the jurisdiction in which they operate, and neither can it prevent sperm donors from donating elsewhere during their lives. There is an acute shortage of sperm donors in many parts of the world and there is obvious pressure from many quarters for donor sperm from those willing and able to provide it to be made available as safely and as freely as possible.

Services

Storage

The sperm is stored in small vials or straws of holding between 0.4 and 1.0 ml and cryogenically preserved in liquid nitrogen tanks. It has been proposed that there should be an upper limit on how long frozen sperm can be stored, however a baby has been conceived in the UK using sperm frozen for 21 years. Before freezing, sperm may be prepared so that it can be used for intracervical insemination (ICI), intrauterine insemination (IUI) or for IVF (or assisted reproduction) (ART).

Use

Sperm supplied by a sperm bank may be used where a woman's partner is infertile or where he carries genetic disease. Increasingly, donor sperm is used to achieve a pregnancy where a woman has no male partner, including among lesbian and bisexual mothers-to-be. Sperm from a sperm donor may also be used in surrogacy arrangements and for creating embryos for embryo donation.

Donor sperm may be supplied by the sperm bank directly to the recipient to enable a woman to perform her own artificial insemination which can be carried out using a needle-less syringe or a cervical cap conception device. The cervical cap conception device allows the donor semen to be held in place close to the cervix for between six to eight hours to allow fertilization to take place. Alternatively, donor sperm can be supplied by a sperm bank through a registered medical practitioner who will perform an appropriate method of insemination or IVF treatment using the donor sperm in order for the woman to become pregnant. From a medical perspective, a pregnancy achieved using donor sperm is no different from a pregnancy achieved using partner sperm, and it is also no different from a pregnancy achieved by sexual intercourse.

Sperm banks may supply other sperm banks or a fertility clinic with donor sperm to be used for achieving pregnancies. Sperm banks may also supply sperm for research or educational purposes.

In countries where sperm banks are allowed to operate, the sperm donor will not usually become the legal father of the children he produces as the result of the use of the sperm he donates, but he will be the 'biological father' of such children. In cases of surrogacy involving embryo donation, a form of 'gestational surrogacy', the 'commissioning mother' or the 'commissioning parents' will not be biologically related to the child and may need to go through an adoption procedure.

As with other forms of third party reproduction, the use of donor sperm from a sperm bank gives rise to a number of moral, legal and ethical issues.

Men may also use a sperm bank to store their own sperm for future use particularly where they anticipate travelling to a war zone or having to undergo chemotherapy which might damage the testes.

Selection

Sperm banks make information available about the sperm donors whose donations they hold in the sperm bank to enable customers to select the donor whose sperm they wish to use. This information is often available by way of an online catalogue. A sperm bank will also usually have facilities to help customers to make their choice and they will be able to advise on the suitability of donors for individual donors and their partners.

Where the recipient woman has a partner she may prefer to use sperm from a donor whose physical features are similar to those of her partner. In many cases, the choice of a donor with the correct blood group will be paramount with particular considerations involving the use of sperm from donors with negative blood groups.

If a surrogate is to be used, such as where the customer is not intending to carry the child, considerations about her blood group etc will also need to be taken into account. Information made available by a sperm bank will usually include the race, height, weight, blood group, health and eye colour of the donor. Sometimes information about his age, family history and educational achievements will also be given.

Some sperm banks make a 'personal profile' of a donor available and occasionally more information may be purchased about a donor, either in the form of a DVD or in written form. Catalogues usually state whether samples supplied in respect of a particular donor have already given rise to pregnancies, but this is not necessarily a guide to the fecundity of the sperm since a donor may not have been in the programme long enough for any pregnancies to have been recorded.

If a woman intends to have more than one child, she may wish to have the additional child or children by the same donor. Sperm banks will usually advise whether sufficient stocks of sperm are available from a particular donor for subsequent pregnancies, and they normally have facilities available so that the woman may purchase and store additional vials from that donor on payment of an appropriate fee. These will be stored until required for subsequent pregnancies or they may be onsold if they become surplus to the woman's requirements.

The catalogue will also state whether samples of sperm are available for Intracervical Insemination ICI, IUI Intrauterine Insemination) or IVF use. ICI is for vaginal or cervical insemination and IUI are 'washed' samples prepared for intrauterine insemination where sperm is injected directly into the uterus. IVF treatment refers to 'in vitro' use, i.e. where fertilization takes place outside the body, literally 'in glass', and therefore to treatments involving donor sperm where an egg is removed from the body and an embryo is created which is then re-inserted, or inserted, into a woman. The preparation for IUI is normally done by a gradient centrifugation where prostaglandins and bacteria is removed from the plasma. Sperm prepared for IVF use normally contains a smaller sample of highly concentrated sperm.

Gender Selection

Some sperm banks enable recipients to choose the gender of their child. Although the methods used do not guarantee 100% success, the chances of being able to select the gender of a child are held to be considerably increased. One of the processes used is the 'swim up' method, whereby a sperm extender is added to the donor's freshly ejaculated sperm and the test-tube is left to settle. After about half-an-hour, the lighter Y chromosome sperm will have swum to the top, leaving the heavier 'x' chromosome sperm at the bottom, thus allowing selection and storage according to gender.

The alternative process is the Percoll Method which is similar to the 'swim up' method but involves additionally the centrifuging of the sperm in a similar way to the washing of samples produced for IUI inseminations, or for IVF purposes. Gender selection is not permitted in a number of countries, including the UK.

Regulation

In the United States sperm banks are regulated by the FDA with new guidelines in effect May 25, 2005. There are also regulation in different states including New York and California. In the EU the sperm banks are regulated by the EU Tissue Directive in effect April 7, 2006.

Selection and Screening of Donors

Using anonymous donor sperm or sperm from a non-anonymous, also called known or identifiable sperm donor through a sperm bank is a safe and reliable method of achieving a pregnancy.

A sperm bank takes a number of steps to ensure the health and quality of the sperm which it supplies and it will inform customers of the checks which it undertakes, providing relevant information about individual donors. A sperm bank will usually guarantee the quality and number of motile sperm available in a sample after thawing. They will try to select men as donors who are particularly fertile and whose sperm will survive the freezing and thawing process. Samples are often sold as containing a particular number of motile sperm per millilitre, and different types of sample may be sold by a sperm bank for differing types of use, e.g. ICI or IUI.

A sperm bank recruits donors via advertising, often in colleges and in local newspapers, and also via the internet. A donor must be a fit healthy male, normally between 18 to 45 years of age, who is willing to undergo frequent and rigorous testing and who is willing to donate his sperm so that it can be used to impregnate women who are unrelated to, and unknown by, him.

The donor must agree to relinquish all legal rights to all children which result from his donations. Where a man donates his sperm to a sperm bank, he will usually not be regarded as the legal father of children which are produced from his donations. The donor must produce his sperm at the sperm bank thus enabling the identity of the donor, once proven, always to be ascertained, and also enabling fresh samples of sperm to be produced for immediate processing.

Sperm is produced by way of masturbation. The sperm will be checked to ensure its fecundity and also to ensure that motile sperm will survive the freezing process. If a man is accepted onto the sperm bank's program as a sperm donor, his sperm will be constantly monitored, the donor will be regularly checked for infectious disease, and samples of his blood will be taken at regular intervals.

A sperm bank may provide a donor with dietary supplements containing herbal or mineral substances such as maca, zinc, vitamin E and argenine which are designed to improve the quality and quantity of the donor's semen, as well as reducing the refractory time (i.e. the time between viable ejaculations). All sperm is frozen in straws or vials and stored for a minimum of 6 months before being released for sale and use to ensure that the donor remains healthy.

Donors are subject to tests for infectious diseases such as human immunoviruses HIV (HIV-1 and HIV-2), human T-cell lymphotropic viruses (HTLV-1 and HTLV-2), syphilis, chlamydia, gonorrhea, Hepatitis B virus, Hepatitis C virus, cytomegalovirus (CMV), Trypanosoma cruzi and Malaria as well as hereditary diseases such as cystic fibrosis, Sickle cell anemia, Familial Mediterranean fever, Gaucher's disease, Thalassaemia, Tay-Sachs disease, Canavan's disease, Familial dysautonomia, Congenital adrenal hyperplasia Carnitine transporter deficiency and Karyotyping 46XY. Karyotyping is not a requirement in neither EU nor the U.S. but some sperm banks choose to test donors as an extra service to the customer.

A sperm donor may also be required to produce his medical records and those of his family, often for several generations. A sperm sample is usually tested micro-biologically at the sperm bank before it is prepared for freezing and subsequent use. A sperm donor's blood group may also be registered to ensure compatibility with the recipient.

Surrogacy

Surrogacy is an arrangement in which a woman carries and delivers a child for another couple or person. This woman may be the child's genetic mother (called traditional surrogacy), or she may carry the pregnancy to delivery after having an embryo, to which she has no genetic relationship whatsoever, transferred to her uterus (called gestational surrogacy). If the pregnant woman received compensation for carrying and delivering the child (besides medical and other reasonable expenses) the arrangement is called a commercial surrogacy, otherwise the arrangement is sometimes referred to as an altruistic surrogacy.

In a traditional surrogacy the child may be conceived via home artificial insemination using fresh or frozen sperm or impregnated via IUI (intrauterine insemination), or ICI (intracervical insemination) performed at a health clinic.

The social parents (that is, those that intend to raise the child) may arrange a surrogate pregnancy because of homosexuality, female infertility, or other medical issues which may make the pregnancy or delivery impossible, risky or otherwise undesirable. The social mother could also be fertile and healthy, and prefer the convenience of someone else undergoing pregnancy, labor, and delivery for her. The intended parent could also be a single man or woman wishing to have his/her own biological child. In Nebraska, involuntarily childless couples are encouraged to adopt instead of looking for a surrogate. However, this idea is not always a practical one—adoption is a long, expensive process. Adoptions typically require the individuals to be married, of a certain age, and sometimes married a certain number of years before they are even considered for adoption.

Some agencies may prohibit adoptions based on disabilities and sexual orientations, and also may require one of the new adoptive parents to stay home with the adopted child for a certain amount of time following the adoption. It is encouraged for the woman wanting to be a surrogate mother to be financially stable. This would prevent woman wanting to become surrogates due to financial need and also would minimise possible exploitation. However, it is historically unlikely that prospective parents would hastily choose a possible surrogate who was not financially reliable.

A large controversy is the impact of a future relationship that the surrogate might have the child after it is born. It is a cultural assumption that normal women do not become pregnant with the premise of being reimbursed monetarily, and that women naturally develop a bond with the child they give birth to. The legality of surrogacy arrangements vary widely between jurisdictions.

History

Having another woman bear a child for a couple to raise, usually with the male half of the couple as the genetic father, is referred to in antiquity. Babylonian law and custom allowed this practice and infertile woman could use the practice to avoid the divorce which would otherwise be inevitable.

In the United States, the issue of surrogacy was widely publicized in the case of Baby M, in which the surrogate and biological mother of Melissa Stern ("Baby M"), born in 1986, refused to cede custody of Melissa to the couple with whom she had made the surrogacy agreement. The courts of New Jersey found that Mary Beth Whitehead was the child's legal mother and declared contracts for surrogate motherhood illegal and invalid.

However, the court found it in the best interests of the infant to award custody of Melissa to her biological father William Stern and his wife Elizabeth Stern, rather than to the surrogate mother Mary Beth Whitehead.

There have been cases of clashes between surrogate mothers and the genetic parents; when unexpected complications with the fetus makes the genetic parents ask for an abortion even though the surrogate mother is opposing the abortion.

Legality

The legal aspects surrounding surrogacy are very complex and mostly unsettled. There is a default legal assumption in most countries that the woman giving birth to a child is that child's legal mother. In some jurisdictions the possibility of surrogacy has been legally allowed and as a result, the intended parents may be recognised as the legal parents right from the birth of a baby. Many states now issue pre-birth orders through the courts placing the name(s) of the intended parent(s) on the birth certificate from the start. In other states that do not issue such orders, the possibility of surrogacy is either not recognized (all contracts specifying different legal parents are void), or is prohibited.

AUSTRALIA

In all states in Australia (except Tasmania which bans all surrogacy under the *Surrogacy Contracts Act 1993*), altruistic surrogacy has been the only recently recognised surrogacy that has become legal. However, in all states and the Australian Capital Territory arranging commercial surrogacy is a criminal offense, although the Northern Territory has no legislation governing surrogacy at all and there are no plans to introduce laws on surrogacy into the NT Legislative Assembly in near the future. In 2006, Australian senator Stephen Conroy and his wife Paula Benson announced that they had arranged for a child to be born through egg donation and gestational surrogacy. Unusually, Conroy was put on the birth certificate as the father of the child. Previously, couples who used to make surrogacy arrangements in Australia had to adopt the child after it was registered as born to the natural mother; rather than being recognised as birth parents, however now that surrogacy is more regular practice for childless parents; most states have switched to such arrangements to give the intended parents proper rights. After the announcement, Victoria changed their legislation since 1 January 2010, under the *Assisted Reproductive Treatment Act 2008* to make altruistic surrogacy within the state legal, however commercial surrogacy is illegal.

Since 1 June 2010 in Queensland, altruistic surrogacy is legal under the *Surrogacy Act 2010 No 2.* Yet, the commercial surrogacy is still illegal under the legislation.

Similarly, in both New South Wales and the Australian Capital Territory, altruistic surrogacy is legal under the *Surrogacy Act 2010 No 102* and the *Parentage Act 2004*, respectively.

In Western Australia (under the *Surrogacy Act 2008*) and South Australia (under the *Family Relationships Act 1975*) altruistic surrogacy is only legal for couples consisting of the opposite sex (single people and same sex couples are banned from altruistic surrogacy). In 2011, Tasmania lawmakers after a review of the laws and a community consultation process introduced the *Surrogacy Bill 2011* and the *Surrogacy (Consequential Amendments) Bill 2011* to the lower house and passed the lower house by a vote of 22-3. Now it is yet to pass the conservative upper house of 15 members (12 of them being independent members, 2 Labor and 1 Liberal).

CANADA

Canada allows surrogacy and it is governed by the Assisted Human Reproduction Act - and it outlines that all expenses that are contributed to the surrogacy can be paid to the surrogate by the intended parents. Expenses must be accompanied with receipts.

QUEBEC

In the province of Quebec, contracts that involve surrogacy are unenforceable.

FRANCE

In France, since 1994, any surrogacy arrangement that is commercial or altruistic, is illegal or unlawful and is not sanctioned by the law (art 16-7 of the Code Civil). The French Courts the Cessation already took this point of view in 1991. It held that if any couple makes an agreement or arranges with another person that she is to bear the husband's child and surrender it on birth to the couple, and that she is choosing that she will not keep the child, the couple making such an agreement or arrangement, is not allowed to adopt the child.

GEORGIA (COUNTRY)

Surrogacy, along with ovum and sperm donation, has been legal in Georgia since 1992. Under applicable law, a donor or surrogate mother has no parental rights over the child born.

HONG KONG

Commercial surrogacy is criminal under the Human Reproductive Technology Ordinance 2000. The law is phrased in a manner that no one can pay a surrogate, no surrogate can receive money, and no one can arrange a commercial surrogacy (the same applies to the supply of gametes), no matter within or outside Hong Kong. Normally only the gametes of the intended parents can be used.

In October 2010, Peter Lee, the eldest son and one of the presumed heirs of billionnaire Lee Shau Kee obtained three sons through a surrogate mother, reportedly from California. Since the junior Lee is single, the news attracted criticism on both moral and legal grounds. A vicar general of the territory's Roman Catholic diocese was critical. In December the case was reportedly referred to police after questions were asked in Legco.

HUNGARY

Commercial surrogacy is illegal in Hungary.

ICELAND

All surrogacy arrangements (both commercial and altruistic) are illegal.

INDIA

Commercial surrogacy is legal in India, as recognized by the Supreme Court of India in 2002. India is emerging as a leader in international surrogacy and a destination in surrogacy-related fertility tourism. Indian surrogates have been increasingly popular with fertile couples in industrialised nations because of the relatively low cost. Indian clinics are at the same time becoming more competitive, not just in the pricing, but in the hiring and retention of Indian females as surrogates.

Clinics charge patients between $10,000 and $28,000 for the complete package, including fertilization, the surrogate's fee, and delivery of the baby at a hospital. Including the costs of flight tickets, medical procedures and hotels, it comes to roughly a third of the price compared with going through the procedure in the UK. The Honourable Supreme Court of India has given the verdict that the citizenship of the child born through this process will have the citizenship of its surrogate mother.

ISRAEL

In March 1996, the Israeli government legalised gestational surrogacy under the "Embryo Carrying Agreements Law." This law made Israel the first country in the world to implement a form of state-controlled surrogacy in which each and every contract must be approved directly by the state.

A state-appointed committee permits surrogacy arrangements to be filed only by Israeli citizens who share the same religion. Surrogates must be single, widowed or divorced and only infertile heterosexual couples are allowed to hire surrogates. The numerous restrictions on surrogacy under Israeli law have prompted some intended parents to turn to surrogates outside of the country.

ITALY

All surrogacy arrangements (both commercial and altruistic) are illegal.

JAPAN

In March 2008, the Science Council of Japan proposed a ban on surrogacy and said that doctors, agents and their clients should be punished for commercial surrogacy arrangements.

NETHERLANDS AND BELGIUM

Altrustic surrogacy is legal in Belgium and the Netherlands. Only commercial surrogacy is illegal in Belgium and the Netherlands. Although altrustic surrogacy is legal, there is only one hospital taking in couples and there are extremely strict rules to get in. This makes a lot of couples seek their treatment outside the Netherlands or Belgium.

PAKISTAN

In Karachi and Lahore, many hospitals are doing this business, government of Pakistan has not yet defined any policy on Surrogacy.

RUSSIAN FEDERATION

Gestational surrogacy, even commercial (is legal in Russia), being (available for practically all adults willing to be parents). There has to be a certain medical indication for surrogacy: absence of uterus; uterine cavity or cervix deformity; uterine cavity synechia; somatic diseases contraindicating child bearing; repeatedly failed IVF attempts, when high-quality embryos were repeatedly obtained and their transfer wasn't followed by pregnancy. The first surrogacy program in Russia was successfully implemented in 1995 at the IVF centre of the Obstetrics and Gynecology Institute in St. Petersburg.

The public opinion is surrogacy-friendly, recent cases of a famous singer and a well-known business-woman, who openly used services of gestational surrogates received very positive news coverage.

A few Russian women such as Ekaterina Zakharova, Natalija Klimova, Lamara Kelesheva became grandmothers through postmortem gestational surrogacy programs, their surrogate grandsons being conceived posthumously after the death of their sons.

Registration of children born through surrogacy is regulated by the Family Code of Russia and the Law on Acts on Civil Status. A surrogate's consent is needed for that.

Apart from that consent, no adoption nor court decision is required. The surrogate's name is never listed on the birth certificate. There is no requirement for the child to be genetically related to at least one of the commissioning parents.

Children born to heterosexual couples who are not officially married or single intended parents through gestational surrogacy are registered in accordance to analogy of jus.

A court decision might be needed for that. On August 5, 2009 a St. Petersburg court definitely resolved a dispute whether single women could apply for surrogacy and obliged the State Registration Authority to register a 35 year old single intended mother Nataliya Gorskaya as the mother of her "surrogate" son.

On 4 August 2010, a Moscow court ruled that a single man who applied for gestational surrogacy (using donor eggs) could be registered as the only parent of his son, becoming the first man in Russia to defend his right to become a father through a court procedure. The surrogate mother's name was not listed on the birth certificate, the father was listed as the only parent.

After that a few more identical decisions concerning single men who became fathers through surrogacy were adopted by different courts in Russia listing men as the only parents of their "surrogate" children and confirming that prospective single parents, regardless of their sex or sexual orientation, can exercise their right to parenthood through surrogacy in Russia.

Liberal legislation makes Russia attractive for "reproductive tourists" looking for techniques not available in their countries. Intended parents come there for oocyte donation, because of advanced age or marital status (single women and single men) and when surrogacy is considered. Foreigners have the same rights as for assisted reproduction as Russian citizens. Within 3 days after the birth the commissioning parents obtain a Russian birth certificate with both their names on it. Genetic relation to the child (in case of donation) just doesn't matter.

SAUDI ARABIA

Religious authorities in Saudi Arabia do not allow the use of surrogate mothers. They have instead suggested medical procedures to restore female fertility and ability to deliver. To this end, Saudi authorities sanctioned the world's first uterus transplant in an infertile woman.

SWEDEN

Surrogacy is not clearly regulated in Swedish law. The legal procedure most equivalent to it is making an adoption of the child from the surrogate mother. However, the surrogate mother thereby has the right to keep the child if she changes her mind until the adoption. Yet, the biological father may claim right to the child.

It is illegal for Swedish fertility clinics to make surrogate arrangements.

UKRAINE

Since 2002, surrogacy and surrogacy in combination with egg/sperm donation has been absolutely legal in Ukraine. According to the law a donor or a surrogate mother has no parental rights over the child born and the child born is legally the child of the prospective parents.

In Ukraine the start of introduction of methods of supporting reproductive medicine was given in eighties of the preceding century. It was Kharkov where the extracorporeal fertilization method was for the first time successfully applied in Ukraine, and in 1991 a girl named Katy was born. Kharkiv was also the first city in CIS countries to realise surrogacy. Many clinics dealing with surrogacy have been opened in Kiev and Lviv.

Ukrainian surrogacy laws are very favourable and fully support the individual's reproductive rights. Surrogacy is officially regulated by Clause 123 of the Family Code of Ukraine and Order 771 of the Health Ministry of Ukraine.

You can choose between Gestational Surrogacy, Egg/sperm Donation, special Embryo adoption programs and their combinations. No specific permission from any regulatory body is required for that. A written informed consent of all parties (intended parents and surrogate) participating in the surrogacy program is mandatory.

Ukrainian legislation allows intended parents to carry on a surrogacy program and their names will be on birth certificate of the child born as a result of the surrogacy programme from the very beginning. The child is considered to be legally "belonging" to the prospective parents from the very moment of conception. The surrogate's name is never listed on the birth certificate.

The surrogate cannot keep the child after the birth. Even if a donation programme took place and there is no biological relation between the child and intended parents, their names will be on birth certificate.

Embryo research is also allowed, gamete and embryo donation permitted on a commercial level. Single women can be treated by known or anonymous donor insemination. Gestational surrogacy is an option for officially married couples and single women. There is no such concept as gay/lesbian marriage in Ukraine, meanwhile such patients can be treated as single women/men.

UNITED KINGDOM

Commercial surrogacy arrangements are not legal in the United Kingdom. Such arrangements were prohibited by the Surrogacy Arrangements Act 1985. Whilst it is illegal in the UK to pay more than expenses for a surrogacy, the relationship is recognised under section 30 of the Human Fertilization and Embryology Act 1990. Regardless of contractual or financial consideration for expenses, surrogacy arrangements are not legally enforceable within the United Kingdom. A surrogate mother still maintains the legal right of determination for the child, even if they are genetically unrelated. Unless a parental order or adoption order is made the surrogate mother remains the legal mother of the child.

UNITED STATES

Many states have their own state laws written regarding the legality of surrogate parenting. It is most common for surrogates to reside in Florida and California due to the surrogacy-accommodating laws in these states. California is especially popular due to its enforceable surrogacy agreements. Surrogacy is well developed around Camp Pendleton in California.

With the accommodating laws of the State of California and the long overseas deployments of husbands, wives have found surrogacy to be a means to supplement military incomes and to provide a needed service. It is illegal to hire a surrogate in New York, and even embryonic transfers may not be done in New York.

At this point, the laws surrounding surrogacy are well defined in the Commonwealth of Pennsylvania, and surrogacy is beginning to become common in the state of Delaware. Attorney Noel Keane and Warren Ringold of Dearborn, Michigan advocated for the passage of laws that protected the idea of surrogate motherhood. Bill Handel, who is a partner in a Los Angeles Surrogacy firm, also attempted to have such laws passed in California, but his attempts were struck down in the State Congress. Presently, the idea of surrogate motherhood has gained some societal acceptance and laws protecting the contractual arrangements exist in eight states.

Ethical Issues

According to Professor Bryan Caplan, libertarians have since Voltaire emphasized the fact that it is not right to forbid people from doing things that others do not accept. However, according to Caplan also non-liberals should support paid surrogacy: creating a human life is almost always good and voluntary exchange is usually beneficial for all participants, also in surrogacy.

Mother-child Relationship

A study by the Family and Child Psychology Research Centre at City University, London, UK in 2002 concluded that surrogate mothers rarely had difficulty relinquishing rights to a surrogate child and that the intended mothers showed greater warmth to the child than mothers conceiving naturally. Anthropological studies of surrogates have shown that surrogates engage in various distancing techniques throughout the surrogate pregnancy so as to ensure that they do not become emotionally attached to the baby. Many surrogates intentionally try to foster the development of emotional attachment between the intended mother and the surrogate child.

Instead of the popular expectation that surrogates feel traumatized after relinquishment, an overwhelming majority describe feeling empowered by their surrogacy experience. In fact, quantitative and qualitative studies of surrogates over the past twenty years, mostly from a psychological or social work perspective, have confirmed that the majority of surrogates are satisfied with their surrogacy experience, do not experience "bonding" with the child they birth, and feel positively about surrogacy even a decade after the birth. Assessing such studies from a social constructionist perspective reveals that the expectation that surrogates are somehow "different" from the majority of women and that they necessarily suffer as a consequence of relinquishing the child have little basis in reality and are instead based on cultural conventions and gendered assumptions.

Many surrogates form close and intimate relationships with the intended parents. When the greatness of their efforts is acknowledged, they recall their surrogacy experience in the years to come as the most meaningful experience of their lives. Even when the surrogate cases are successful in providing couples with children, there is the possibility of problems in the existing marriage surfacing. The thought of one of the partners in the relationship being unable to have children may lead to the need for counselling and medical support from a doctor or clinician.

In Fiction

Films and other fiction depicting emotional struggles of assisted reproductive technology have had an upswing first in the latter part of the 2000s decade, although the techniques have been available for decades.

Yet, the amount of people that can relate to it by personal experience in one way or another is ever growing, and the variety of trials and struggles is huge.

- In Robert A. Heinlein's 1961 science fiction novel *Stranger in a Strange Land,* TV reporters exult that socialite Cynthia Duchess has decided to have the Perfect Baby, with the use of a sperm donor, an egg donor, and a gestational carrier.
- In the *Metal Gear* video game series, the character EVA was the gestational surrogate mother who gave birth to Solid Snake and Liquid Snake.
- The heroine of Heinlein's 1966 novel *The Moon Is a Harsh Mistress* describes herself as a "professional host-mother".
- The 1986 novel *To Have and To Hold* by Deborah Moggach (and the TV series on which it is based) has surrogacy as its central theme.
- The character Phoebe Buffay in hit US sitcom *Friends* was a gestational surrogate mother for her brother and his wife.
- The character Kira Nerys in sci-fi show *Star Trek: Deep Space Nine* acted as surrogate for the child of her colleague Miles O'Brien and his wife Keiko, after Keiko was injured and no longer able to carry. This was especially unusual as Kira and the O'Briens are not even the same species. Kira moved in with the O'Briens during the pregnancy. The O'Briens honoured Kira's wish for a traditional Bajoran birth rather than a human one, and even named the child after Kira.
- The character Fertility in Chuck Palahniuk's novel *Survivor* works as a commercial surrogate, although she is infertile herself.
- In the 2008 film *Baby Mama,* starring Tina Fey and Amy Poehler, Amy Poehler's character is a surrogate to Tina Fey's character.

- The Character Leah Patterson/Baker was surrogate to her friend Sally Fletcher and husband Flynn Saunders in the Australian soap opera *Home and Away.*
- In the film Chori Chori Chupke Chupke Madhu becomes a surrogate mother for Raj Malhotra and his wife Priya after Priya suffers an injury that causes her to become infertile.
- The character Christina McKinney in *Ugly Betty* acts as a surrogate for Wilhelmina Slater.
- The character Morgan Braithwaite in New Zealand's Serial Drama, Shortland Street, is an illegal surrogate for her infertile friends, Cindy and Trent.
- The character Ganondorf from The Legend of Zelda series is the surrogate child of Twinrova.
- The character Rachel Berry from Glee was born through a surrogate mother, played by Idina Menzel, to two gay fathers.
- The character Eliza Makepeace in Kate Morton's novel *The Forgotten Garden* is a surrogate mother for her cousin, Rose, who had her ovaries damaged by overlong exposure to X-rays in her childhood.

Transgenic Vegetables

When speaking about transgenic foodstuff we refer to those vegetables and cereals which have been genetically modified so as to obtain a bigger production. Among these products we could mention certain varieties of corn, soy bean, cotton, tomatoes and potatoes.

Some research has been done also for transgenic cultivation of melons, plums, tobacco... and even for animals such as transgenic pigs. Transgenic living specimens are obtained by introducing new genetic information into the chromosomes of the species. The procedure is similar to the one followed by some varieties of viruses which contaminate the cells and cause them to generate new viruses by means of introducing certain molecules in them.

In fact, one of the most common technics for plants consists in the use of genetic substance of a virus or bacterium to obtain the incorporation of the new genetic code in the host organism.

The Genes

Genetically modified vegetables contain as many as three different genes: one offering resistance to antibiotics, a self-insecticide one and a thrid one offering resistance to herbicides.

The *antibiotic resisting gene* is a marking gene which in the laboratory process of selection permits to differentiate between those individuals which have been modified and those which have not. This gene traditionally incorporates information of resistance to ampiciline, although other antibiotics for use on humans are used for that also.

At present these technics are being modified by using as a marker an antibiotic existing in nature and not being a pharmaceutical element.

The *self-insecticide gene* is the origin of transgenic vegetables. Information to generate a so called Bt toxin is incorporated in plants. This toxin gets rid of insects harmful to modified cultivation.

Formerly it was used as an insecticide, the dosing process being the only difference. The *gene resistant to herbicides* is sometimes used as a marker or also as an increased quality of the cultivation itself.

This allows for the use of a bigger amount of herbicide in order to get rid of plagues in such a way that the cultivated vegetable be not affected.

Late research by one of the specialised multinational firms -Monsanto Co.- is trying to develop new genetic technics called *'terminator'* seeking to obtain sterile genetic modified vegetables or sterile seeds. As a result the impact on biodiversity would diminish.

If, in spite of the antagonism which such a research meets with everywhere, these technics eventually meet with success they will become the fourth genetic modification incorporated in transgenic foodstuff.

The Advantages of Transgenic Vegetables

The promotion of transgenic vegetables by this branch of multinational firms rests fundamentally on a twofold conception: a larger productivity and the possibility of doing away with hunger in the world.

A larger productivity: Incorporating insecticide in the plant itself brings forth a greater efficiency in the struggle against specific parasites, such as corn deathwatch, which are not beaten efficaciously by means of external distribution of insecticides. This yields a bigger production as a result. At present the experience of USA farmers place the increase in productivity at a 9%.

More ecological: These new insecticide distribution technics allow for a much more localised application and prevents dispersion in the environment.

Doing away with hunger in the world: One of the slogans set forth by transgenic vegetable producer firms states that doing away with hunger in the world is possible. Needless to say that the problem of hunger does not lie in the lack of resources but in their distribution.

The Hazards of Transgenic Foodstuff

Here are some of the possible objections to transgenic foodstuff:

Lack of information and transparency: Some transgenic vegetables, such as soy bean and corn, are used mixed up indiscriminately with non-modified cereals. This, together with the absence of a legislation on control and labelling of these new products, leads to the fact that consumers have no rights to either being conscious of what they are consuming or choosing between consuming or not consuming transgenic foodstuff. Besides, transgenic foodstuff is used to feed cattle and fowl. Knowing how the animals we consume were fed is next to impossible.

Lack of health prevention: Introducing a new composite into a cell - such is the case of plasmodium including a new genetic information- could alter the metabolic stability of host vegetable cells thus generating new composites or altering the nutritious contents of modified foodstuff.

Such alterations have been observed in genetic engineering experiments.

As a result genetic alteration holds in its effects a certain incidental condition which conventional foodstuff does not and which should be restrained.

Absence of ethic and political justifiability: Another set of objections to transgenic foodstuff points to the needlessness and hence the lack of ethic feelings towards genetic manipulation.

On the other hand, certain political positions point out that there is a greater concern in the interests of some firms than in the well-being of society.

Environmental hazards: There are more and more perceptions giving to understand that genetic engineering implies a new liability to environment, especially in so far as the effects on biodiversity and genetic contamination are concerned.

A diminishing amount of cultivated species and the expansion of toxic species for some organisms are examples of liability to the maintenance of biodiversity.

Moreover, genetic information introduced into transgenic vegetables has a remote possibility of migrating and installing itself in other organisms as if it were a virus. This may cause the presence of bacteria resistant to the antibiotics used in the selection of modified organisms, as well as new plagues resistant to the same insecticides or herbicides, or both, as the cultivated plants.

One last notion has to do with the sense of a gene resistant to herbicides. It permits the increase in dose of herbicides in crops so that the supply of matter alien to environment (pollution) increases.

Economic Control: On the lines laid down by producer firms some procedures of transgenic seed producion may lead to a new dependency of the South from the North and of the entire world from such multinational firms:

- Doubtful procedure with regard to registration of trade marks.
- Acquisition of conventional seed producer firms so as to shut them out from marketing or make use of them as distributors of modified seeds.
- Binding seed purchasers to signing contracts including prohibition to keep in reserve part of the seeds, the obligation to using herbicides from the same commercial firm, etc.
- Research in technology of sterilizing seeds so as to make their conservation impossible and make it necessary to buy fresh seeds every year.

6

Xenotransplantation

Xenotransplantation (*xenos*- from the Greek meaning “foreign”) is the transplantation of living cells, tissues or organs from one species to another. Such cells, tissues or organs are called xenografts or xenotransplants. In contrast, the term allotransplantation refers to a same-species transplant. Human xenotransplantation offers a potential treatment for end-stage organ failure, a significant health problem in parts of the industrialised world. It also raises many novel medical, legal and ethical issues.

A continuing concern is that many animals, such as pigs, have shorter lifespans than humans, meaning that their tissues age at a quicker rate. Disease transmission (xenozoonosis) and permanent alteration to the genetic code of animals are also causes for concern. There are few published cases of successful xenotransplantation.

Potential Future Uses

Because there is a worldwide shortage of organs for clinical implantation, about 60% of patients awaiting replacement organs die on the waiting list. Certain procedures, some of which are being investigated in early clinical trials, aim to use cells or tissues from other species to treat life-threatening and debilitating illnesses such as cancer, diabetes, liver failure and Parkinson’s disease. If vitrification can be perfected, it could allow for long-term storage of xenogenic cells, tissues and organs so that they would be more readily available for transplant.

Xenotransplants could save thousands of patients waiting for donated organs. The animal organ, probably from a pig or baboon could be genetically altered with human genes to trick a patient's immune system into accepting it as a part of its own body. They have re-emerged because of the lack of organs available and the constant battle to keep immune systems from rejecting allotransplants. Xenotransplants are thus potentially a more effective alternative.

Also, xenotransplantation of ovarian tissue into immunodeficient nude mice or SCID mice is already used in research to study the development of ovarian follicles. Mature follicles have developed, even after use of cryopreserved ovarian tissue. Both host and graft vessels contribute to the revascularization of xenografted human ovarian tissue in a mice.

Potential Future Animal Organ Donors

Since they are the closest relatives to humans, nonhuman primates were first considered as a potential organ source for xenotransplantation to humans. Chimpanzees were originally considered to be the best option since their organs are of similar size, and they have good blood type compatibility with humans. However, since chimpanzees are listed as an endangered species, other potential donors were sought out.

Baboons are more readily available, however they are also not practical as potential donors. Problems include their smaller body size, the infrequency of blood group O (the universal donor), their long gestation period, and they typically produce few offspring. In addition, a major problem with the use of nonhuman primates is the increased risk of disease transmission, since they are so closely related to humans. Pigs are currently thought to be the best candidates for organ donation. The risk of cross-species disease transmission is decreased because of their increased phylogenetic distance from humans . They are readily available, their organs are anatomically comparable in size, and new infectious agents are less likely since they have been in close contact with humans through domestication for many generations . Current experiments in xenotransplantation most often use pigs as the donor, and baboons as human models.

Barriers and Issues

Immunologic Barriers

To date no xenotransplantation trials have been entirely successful due to the many obstacles arising from the response of the recipient's immune system. This response, which is generally more extreme than in allotransplantations, ultimately results in rejection of the xenograft, and can in some cases result in the immediate death of the recipient. There are several types of rejection organ xenografts are faced with, these include:

- Hyperacute rejection
- Acute vascular rejection
- Cellular rejection
- Chronic rejection.

A rapid, violent hyperacute response results due to preformed natural antibodies, known as XNAs.

Hyperacute Rejection

This rapid and violent type of rejection occurs within minutes to hours from the time of the transplant. It is mediated by the binding of XNAs (xenoreactive natural antibodies) to the donor endothelium, causing activation of the human complement system which results in endothelial damage, inflammation, thrombosis and necrosis of the transplant. XNAs are first produced and begin circulating in the blood in neonates, after colonisation of the bowel by bacteria which have galactose moieties on their cell walls. Most of these antibodies are the IgM class, but also include IgG, and IgA. .

The epitope XNAs target is an α-linked galactose moiety, Gal-α-1,3Gal (also called the α-Gal epitope), produced by the enzyme α-galactosyl transferase. Most non-primates contain this enzyme thus, this epitope is present on the organ epithelium and is perceived as a foreign antigen by primates, which lack the galactosyl transferase enzyme. In pig to primate xenotransplantation, XNAs recognise porcine glycoproteins of the integrin family .

The binding of XNAs initiate complement activation through the classical complement pathway. Complement activation causes a cascade of events leading to: destruction of endothelial cells, platlet degranulation, inflammation, coagulation, fibrin deposition, and hemorrhage. The end result is thrombosis and necrosis of the xenograft.

Overcoming Hyperacute Rejection

Since hyperacute rejection presents such a barrier to the success of xenografts several strategies to overcome it are under investigation:

Interruption of the complement cascade:

- The recipient's complement cascade can be inhibited through the use of cobra venom factor (which depletes C3), soluble complement receptor type 1, anti-C5 antibodies, or C1 inhibitor (C1-INH). Disadvantages of this approach include the toxicity of cobra venom factor, and most importantly these treatments would deprive the individual of a functional complement system.

Transgeneic organs (Genetically engineered pigs):

- 1,3 galactosyl transferase gene knockouts - These pigs do not contain the gene which codes for the enzyme responsible for expression of the immunogeneic gal-α-1,3Gal moiety (the α-Gal epitope).
- Increased expression of H-transferase (α 1,2 fucosyltransferase), an enzyme that competes with galactosyl transferase. Experiments have shown this reduces α-Gal expression by 70%.
- Expression of human complement regulators (CD55, CD46, and CD59) to inhibit the complement cascade.
- Plasmaphoresis, on humans to remove 1,3 galactosyl-transferase, reduces the risk of activation of effector cells such as CTL (CD8 T cells), complement pathway activation and delayed type hypersensitivity (DTH).

Acute Vascular Rejection

Also known as delayed xenoactive rejection, this type of rejection occurs in discordant xenografts within 2 to 3 days, if hyperacute rejection is prevented. The process is much more complex than hyperacute rejection and is currently not completely understood. Acute vascular rejection requires de novo protein synthesis and is driven by interactions between the graft endothelial cells and host antibodies, macrophages, and platelets.

The response is characterized by an inflammatory infiltrate of mostly macrophages and natural killer cells (with small numbers of T cells), intravascular thrombosis, and fibrinoid necrosis of vessel walls. Binding of the previously mentioned XNAs to the donor endothelium leads to the activation of host macrophages as well as the endothelium itself. The endothelium activation is considered type II since gene induction and protein synthesis are involved.

The binding of XNAs ultimately leads to the development of a procoagulant state, the secretion of inflammatory cytokines and chemokines, as well as expression of leukocyte adhesion molecules such as E-selectin, intercellular adhesion molecule-1 (ICAM-1), and vascular cell adhesion molecule-1 (VCAM-1) . This response is further perpetuated as normally binding between regulatory proteins and their ligands aid in the control of coagulation and inflammatory responses. However, due to molecular incompatibilities between the molecules of the donor species and recipient (such as porcine major histocompatibility complex molecules and human natural killer cells), this may not occur.

Overcoming Acute Vascular Rejection

Due to its complexity, with the use of immunosuppressive drugs along with a wide array of approaches are necessary to prevent acute vascular rejection, and include:

- Administering a synthetic thrombin inhibitor to modulate thrombogenesis
- Depletion of anti-galactose antibodies (XNAs) by techniques such as immunoadsorption, to prevent endothelial cell activation
- Inhibiting activation of macrophages (stimulated by $CD4^+$ T cells) and NK cells (stimulated by the release of Il-2). Thus, the role of MHC molecules and T cell responses in activation would have to be reassessed for each species combo.

Accommodation

If hyperacute and acute vascular rejection are avoided accommodation is possible, which is the survival of the xenograft despite the presence of circulating XNAs. The graft is given a break from humoral rejection when the complement cascade is interrupted, circulating antibodies are removed, or their function is changed, or there is a change in the expression of surface antigens on the graft. This allows the xenograft to upregulate and express protective genes, which aid in resistance to injury, such as heme oxygenase-1 (an enzyme that catalyses the degradation of heme).

Cellular Rejection

Rejection of the xenograft in hyperactute and acute vascular rejection is due to the response of the humoral immune system, since the response is elicited by the XNAs. Cellular rejection is based on cellular immunity, and is mediated by:

1. Natural killer cells, which accumulate in and damagee the xenograft; and
2. T-lymphocytes - which are activated by MHC molecules through both direct and indirect xenorecognition.

In direct xenorecognition, antigen presenting cells from the xenograft present peptides to recipient $CD4^+$ T cells via xenogeneic MHC class II molecules, resulting in the production of interleukin 2 (IL-2). Indirect xenorecognition involves the presentation of antigens from the xenograft by recipient antigen presenting cells to $CD4^+$ T cells. Antigens of phagocytosed graft cells can also be presented by the host's class I MHC molecules to $CD8^+$ T cells. The strength of cellular rejection in xenografts remains uncertain, however it is expected to be stronger than in allografts due to differences in peptides among different animals. This leads to more antigens potentially recognised as foreign, thus eliciting a greater indirect xenogenic response .

Overcoming Cellular Rejection

A proposed strategy to avoid cellular rejection is to induce donor non-responsiveness using haematopoietic chimerism. Donor stem cells are introduced into the bone marrow of the recipient, where they coexist with the recipient's stem cells.

The bone marrow stem cells give rise to cells of all haematopoietic lineages, through the process of hematopoiesis. Lymphoid progenitor cells are created by this process and move to the thymus where negative selection eliminates T cells found to be reactive to self.

The existence of donor stem cells in the recipient's bone marrow causes donor reactive T-cells to be considered self and undergo apoptosis .

Chronic Rejection

This final type of rejection is slow and progressive, and is usually described in transplants which survive the initial rejection phases. Scientists are still unclear how chronic rejection exactly works, research in this area is difficult since xenografts rarely survive past the initial acute rejection phases.

Nonetheless, it is known is that XNAs and the complement system are not primarily involved. Fibrosis in the xenograft occurs as a result of immune reactions, cytokines (which stimulate fibroblasts), or healing (following cellular necrosis in acute rejection). Perhaps the major cause of chronic rejection is arteriosclerosis. Lymphocytes, which were previously activated by antigens in the vessel wall of the graft, activate macrophages to secrete smooth muscle growth factors.

This results in a build up of smooth muscle cells on the vessel walls, causing the hardening and narrowing of vessels within the graft. Chronic rejection leads to pathologic changes of the organ, and is why transplants must be replaced after so many years.

It is also anticipated that chronic rejection will be more aggressive in xenotransplants as opposed to allotransplants.

Physiology

Extensive research is required to determine whether animal organs can replace the physiological functions of human organs.

Many issues include:

- Size - Differences in organ size limit the range of potential recipients of xenotransplants.
- Longevity - The lifespan of most pigs is roughly 15 years, currently it is unknown whether or not a xenograft may be able to last longer than that.
- Hormone and protein differences - Some proteins will be molecularly incompatible, which could cause malfunction of important regulatory processes. These differences also make the prospect of hepatic xenotransplantation less promising, since the liver plays an important role in the production of so many proteins .
- Environment - For example, pig hearts work in a different anatomical site and under different hydrostatic pressure than in humans.
- Temperature - The body temperature of pigs is 39°C (2°C above the average human body temperature). Implications of this difference, if any, on the activity of important enzymes are currently unknown.

Xenozoonosis

Xenozoonosis, also known as zoonosis or xenosis, is the transmission of infectious agents between species via a xenograft. Animal to human infection is normally rare, but has occurred in the past. An example of such is the avian influenza, when an influenza A virus was passed from birds to humans. Xenotransplantation may increase the chance of disease transmission for 3 reasons: (1) Implantation breaches the physical barrier that normally helps to prevent disease transmission, (2) The recipient of the transplant will be severely immunosuppressed; and (3). Human complement regulators (CD46, CD55, and CD59) expressed in transgenic pigs have been shown to serve as virus receptors, and may also help to protect viruses from attack by the complement system.

Examples of viruses carried by pigs include porcine herpesvirus, rotavirus, parvovirus, and circovirus. Porcine herpesviruses and rotaviruses can be eliminated from the donor pool by screening, however others (such as parvovirus and circovirus) may contaminate food and footwear then re-infect the herd.

Thus, pigs to be used as organ donors will have to be housed under strict regulations and screened regularly for microbes and pathogens. Unknown viruses, as well as those which aren't harmful in the animal, may also pose risks (Takeuchi and George, 2000).

Of particular concern are PERVS (porcine endogenous retroviruses), vertically transmitted microbes which are imbedded in swine genomes. The risks with xenosis are twofold as not only could the individual become infected, but a novel infection could initiate an epidemic in the human population.

Because of this risk, the FDA has suggested any recipients of xenotransplants shall be closely monitored for the remainder of their life, and quarantined if they show signs of xenosis.

Baboons and pigs carry myriad transmittable agents which are harmless in their natural host, but extremely toxic and deadly in humans. HIV is an example of a disease which is believed to have jumped from monkeys to humans. Researchers also do not know if an outbreak of infectious diseases could occur and if they could contain the outbreak even though they have measures for control. Another obstacle facing xenotransplants is that of the body's rejection of foreign objects by its immune system. These antigens (foreign objects) are often treated with powerful immunosuppressive drugs which could in turn make the patient vulnerable to other infections and actually aid the disease trying to be cured. This is the reason the organs would have to be altered to fit with the patients' DNA (histocompatibility).

In 2005, the Australian National Health and Medical Research Council (NHMRC) declared a eighteen-year moratorium on all animal-to-human transplantation, concluding that the risks of transmission of animal viruses to patients and the wider community had not been resolved. This was repealed in 2009 after an NHMRC review stated "... the risks, if appropriately regulated, are minimal and acceptable given the potential benefits.", citing international developments on the management and regulation of xenotransplantation by the World Health Organisation and the European Medicines Agency.

Porcine Endogenous Retroviruses

Endogenous retroviruses are remnants of ancient viral infections, found in the genomes of most, if not all, mammalian species. Integrated into the chromosomal DNA, they are vertically transferred through inheritance. Due to the many deletions and mutations they accumulate over time, they usually are not infectious in the host species, however the virus may become infectious in another species . PERVS were originally discovered as retrovirus particles released from cultured porcine kidney cells. Most breeds of swine harbour approximately 50 PERV genomes in their DNA. Although it is likely that most of these are defective, some may be able to produce infectious viruses so every proviral genome must be sequenced to identify which ones pose a threat.

In addition, through complementation and genetic recombination, two defective PERV genomes could give rise to an infectious virus :. There are three subgroups of infectious PERVs (PERV-A, PERV-B, and PERV-C). Experiments have shown that PERV-A and PERV-B can infect human cells in culture. To date no experimental xenotransplantations have demonstrated PERV transmission, yet this does not mean PERV infections in humans are impossible.

Ethicality

Xenografts have been a controversial procedure since they were first attempted. Many, including animal rights groups, strongly oppose killing animals in order to harvest their organs for human use. Religious beliefs, such as the Jewish and Muslim prohibition against eating pork, have been sometimes thought to be a problem, however according to a Council of Europe documentation both religions agree that this rule is overridden by the preservation of human life. Nevertheless, the prohibition of the consumption of pig, does pose problems in some Islamic communities.

In general, the use of pig and cow tissue in humans has been met with little resistance, save some religious beliefs, a few philosophical objections and the occasional satirical barb. With the ability to now use these techniques now manifest, a precise moral and ethical doctrine from all major religions should be pronounced to keep the religious zealots out of the medical technological and testing regime.

Experimentation without consent doctrines are now followed, which was not the case in the past, which may lead to new religious guidelines to further medical research on pronouced ecumenical guidelines. The "Common Rule" is the United States bio-ethics mandate as of 2011. Some ethical issues include informed consent complexities for research subjects, as well as the selection of human subjects, rights of patients and medical staff and public education (as many companies may go ahead with experiments without public awareness).

Split Transplants

Sometimes a deceased-donor organ, usually a liver, may be divided between two recipients, especially an adult and a child. This is not usually a preferred option because the transplantation of a whole organ is more successful.

Domino Transplants

This operation is performed on patients with cystic fibrosis because both lungs need to be replaced and it is a technically easier operation to replace the heart and lungs at the same time. As the recipient's native heart is usually healthy, it can be transplanted into someone else needing a heart transplant. This term is also used for a special form of liver transplant in which the recipient suffers from familial amyloidotic polyneuropathy, a disease where the liver slowly produces a protein that damages other organs.

This patient's liver can be transplanted into an older patient who is likely to die from other causes before a problem arises. This term also refers to a series of living donor transplants in which one donor donates to the highest recipient on the waiting list and the transplant centre utilises that donation to facilitate multiple transplants. These other transplants are otherwise impossible due to blood type or antibody barriers to transplantation. The "Good Samaritan" kidney is transplanted into one of the other recipients, whose donor in turn donates his or her kidney to an unrelated recipient. Depending on the patients on the waiting list, this has sometimes been repeated for up to six pairs, with the final donor donating to the patient at the top of the list. This method allows all organ recipients to get a transplant even if their living donor is not a match to them. This further benefits patients below any of these recipients on waiting lists, as they move closer to the top of the list for a deceased-donor organ. Johns Hopkins Medical Centre in Baltimore and Northwestern University's Northwestern Memorial Hospital have received significant attention for pioneering transplants of this kind.

Major Organs and Tissues Transplanted

Thoracic Organs

- Heart (Deceased-donor only)
- Lung (Deceased-donor and living-related lung transplantation)
- Heart/Lung (Deceased-donor and Domino transplant).

Abdominal Organs

- Kidney (Deceased-donor and Living-Donor)
- Liver (Deceased-donor and Living-Donor)
- Pancreas (Deceased-donor only)
- Intestine (Deceased-donor and Living-Donor)
- Stomach (Deceased-donor only)
- Testis.

Tissues, Cells, Fluids

- Hand (Deceased-donor only),
- Cornea (Deceased-donor only),
- Skin including Face replant (autograft) and Face transplant (extremely rare)
- Islets of Langerhans (Pancreas Islet Cells) (Deceased-donor and Living-Donor)
- Bone marrow/Adult stem cell (Living-Donor and Autograft)
- Blood transfusion/Blood Parts Transfusion (Living-Donor and Autograft)
- Blood vessels (Autograft and Deceased-Donor)

- Heart valve (Deceased-Donor, Living-Donor and Xenograft[Porcine/bovine])
- Bone (Deceased-Donor and Living-Donor).

Types of Donor

Organ donors may be living, or brain dead. Brain dead means the donor must have received an injury (either traumatic or pathological) to the part of the brain that controls heartbeat and breathing. Breathing is maintained via artificial sources, which, in turn, maintains heartbeat. Once brain death has been declared the person can be considered for organ donation. Criteria for brain death vary.

Because less than 3% of all deaths in the U.S. are the result of brain death, the overwhelming majority of deaths are ineligible for organ donation, resulting in severe shortages. Tissue may be recovered from donors who are cardiac dead. That is, their breathing and heartbeat has ceased.

They are referred to as cadaveric donors. In general, tissues may be recovered from donors up to 24 hours past the cessation of heartbeat. In contrast to organs, most tissues (with the exception of corneas) can be preserved and stored for up to five years, meaning they can be "banked." Also, more than 60 grafts may be obtained from a single tissue donor.

Because of these three factors, the ability to recover from a non-heart beating donor, the ability to bank tissue, and the number of grafts available from each donor, tissue transplants are much more common than organ transplants. The American Association of Tissue Banks estimates that more than one million tissue transplants take place in the United States each year.

Living

In "living donors", the donor remains alive and donates a renewable tissue, cell, or fluid (e.g. blood, skin), or donates an organ or part of an organ in which the remaining organ can regenerate or take on the workload of the rest of the organ (primarily single kidney donation, partial donation of liver, small bowel). Regenerative medicine may one day allow for laboratory-grown organs, using patient's own cells via stem cells, or healthy cells extracted from the failing organs.

Deceased

Deceased (formerly cadaveric) are donors who have been declared brain-dead and whose organs are kept viable by ventilators or other mechanical mechanisms until they can be excised for transplantation. Apart from brain-stem dead donors, who have formed the majority of deceased donors for the last twenty years, there is increasing use of Donation after Cardiac Death Donors (formerly non-heart beating donors) to increase the potential pool of donors as demand for transplants continues to grow. These organs have inferior outcomes to organs from a brain-dead donor; however given the scarcity of suitable organs and the number of people who die waiting, any potentially suitable organ must be considered.

Reasons for Donation and Ethical Issues

Living Related Donors

Living related donors donate to family members or friends in whom they have an emotional investment. The risk of surgery is offset by the psychological benefit of not losing someone related to them, or not seeing them suffer the ill effects of waiting on a list.

Paired Exchange

A "paired-exchange" is a technique of matching willing living donors to compatible recipients using serotyping. For example a spouse may be willing to donate a kidney to their partner but cannot since there is not a biological match. The willing spouse's kidney is donated to a matching recipient who also has an incompatible but willing spouse. The second donor must match the first recipient to complete the pair exchange. Typically the surgeries are scheduled simultaneously in case one of the donors decides to back out and the couples are kept anonymous from each other until after the transplant.

Paired exchange program were popularized in the *New England Journal* of Medicine article "Ethics of a paired-kidney-exchange program" in 1997 by L.F. Ross. It was also proposed by Felix T. Rapport in 1986 as part of his initial proposals for live-donor transplants "The case for a living emotionally related international kidney donor exchange registry" in *Transplant Proceedings*. A paired exchange is the simplest case of a much larger exchange registry program where willing donors are matched with any number of compatible recipients. Transplant exchange programs have been suggested as early as 1970: "A cooperative kidney typing and exchange program.".

The first pair exchange transplant in the U.S. was in 2001 at Johns Hopkins Hospital. The first complex multihospital kidney exchange involving 12 patients was performed in February 2009 by The Johns Hopkins Hospital, Barnes-Jewish Hospital in St. Louis and Integris Baptist Medical Centre in Oklahoma City.

Another 12-patient multihospital kidney exchange was performed four weeks later by Saint Barnabas Medical Centre in Livingston, New Jersey,

Newark Beth Israel Medical Centre and New York-Presbyterian Hospital. Surgical teams led by Johns Hopkins continue to pioneer in this field by having more complex chain of exchange such as eight-way multihospital kidney exchange. In December 2009, a 13 organ 13 recipient matched kidney exchange took place, coordinated through Georgetown University Hospital and Washington Hospital Centre, Washington DC.

Paired-donor exchange, led by work in the New England Program for Kidney Exchange as well as at Johns Hopkins University and the Ohio OPOs may more efficiently allocate organs and lead to more transplants.

Good Samaritan

Good Samaritan or "altruistic" donation is giving a donation to someone not well-known to the donor. Some people choose to do this out of a need to donate. Some donate to the next person on the list; others use some method of choosing a recipient based on criteria important to them. Web sites are being developed that facilitate such donation. It has been featured in recent television journalism that over half of the members of the Jesus Christians, an Australian religious group, have donated kidneys in such a fashion.

Compensated Donation

In compensated donation, donors get money or other compensation in exchange for their organs. This practice is common in some parts of the world, whether legal or not, and is one of the many factors driving medical tourism.

In the United States, The National Organ Transplant Act of 1984 made organ sales illegal. In the United Kingdom, the Human Organ Transplants Act 1989 first made organ sales illegal, and has been superseded by the Human Tissue Act 2004.

In 2007, two major European conferences recommended against the sale of organs.

Recent development of web sites and personal advertisements for organs among listed candidates has raised the stakes when it comes to the selling of organs, and have also sparked significant ethical debates over directed donation, "good-Samaritan" donation, and the current U.S. organ allocation policy. Bioethicist Jacob M. Appel has argued that organ solicitation on billboards and the internet may actually increase the overall supply of organs.

Two books, *Kidney for Sale By Owner* by Mark Cherry (Georgetown University Press, 2005); and *Stakes and Kidneys: Why markets in human body parts are morally imperative* by James Stacey Taylor: (Ashgate Press, 2005); advocate using markets to increase the supply of organs available for transplantation. In a 2004 journal article Economist Alex Tabarrok argues that allowing organ sales, and elimination of organ donor lists will increase supply, lower costs and diminish social anxiety towards organ markets. Iran has had a legal market for kidneys since 1988, and the market price is of the order of US$1,200 for the recipient. *The Economist* and the Ayn Rand Institute approve and advocate a legal market elsewhere. They argued that if 0.06% of Americans between 19 and 65 were to sell one kidney, the national waiting list would disappear (which, *the Economist* wrote, happened in Iran). The Economist argued that donating kidneys is no more risky than surrogate motherhood, which can be done legally for pay in most countries.

In Pakistan, 40 percent to 50 percent of the residents of some villages have only one kidney because they have sold the other for a transplant into a wealthy person, probably from another country, said Dr. Farhat Moazam of Pakistan, at a World Health Organization conference. Pakistani donors are offered $2,500 for a kidney but receive only about half of that because middlemen take so much. In Chennai, southern India, poor fishermen and their families sold kidneys after their livelihoods were destroyed by the Indian Ocean tsunami on December 26, 2004. About 100 people, mostly women, sold their kidneys for 40,000–60,000 rupees ($900–$1,350). Thilakavathy Agatheesh, 30, who sold a kidney in May 2005 for 40,000 rupees said, "I used to earn some money selling fish but now the post-surgery stomach cramps prevent me from going to work." Most kidney sellers say that selling their kidney was a mistake.

In Cyprus in 2010 police closed a fertility clinic under charges of trafficking in human eggs. The Petra Clinic, as it was known locally, imported women from Ukraine and Russia for egg harvesting and sold the genetic material to foreign fertility tourists. This sort of reproductive trafficking violates laws in the European Union. In 2010 the Pulitzer Centre on Crisis Reporting and the magazine Fast Company explored illicit fertility networks in Spain, the United States and Israel.

Allocation of Donated Organs

The overwhelming majority of deceased-donor organs in the United States are allocated by federal contract to the Organ Procurement and Transplantation Network (OPTN), held since it was created by the Organ Transplant Act of 1984 by the United Network for Organ Sharing or UNOS. (UNOS does not handle donor cornea tissue; corneal donor tissue is usually handled by various eye banks.) UNOS allocates organs based on the method considered most fair by the scientific leadership in the field. For kidneys, for instance, that is by waiting time; for livers, it is by MELD (Model of End-Stage Liver Disease), an empirical score based on lab values indicative of the sickness of the patient from liver disease.

Experiencing somewhat increased popularity, but still very rare, is directed or targeted donation, in which the family of a deceased donor (often honouring the wishes of the deceased) requests an organ be given to a specific person. If medically suitable, the allocation system is subverted, and the organ is given to that person. In the United States, there are various lengths of waiting due to the different availabilities of organs in different UNOS regions. In other countries such as the UK, only medical factors and the position on the waiting list can affect who receives the organ. If this is not the desired person, it is noted that this puts them higher on the list.

One of the more publicized cases of this type was the 1994 Chester and Patti Szuber transplant. This was the first time that a parent had received a heart donated by one of their own children. Although the decision to accept the heart from their recently killed child was not an easy decision, the Szuber family agreed that giving Patti's heart to her father would have been something that she would have wanted.

Access to organ transplantation is one reason for the growth of medical tourism.

Forced Donation

There have been various accusations that certain authorities are harvesting organs from those the authorities deem undesirable, such as prison populations. The World Medical Association stated that individuals in detention are not in the position to give free consent to donate their organs .

Illegal dissection of corpses is a form of body-snatching and may have taken place to obtain allografts. According to the Chinese Deputy Minister of Health, Huang Jiefu, approximately 95% of all organs used for transplantation are from executed prisoners. The lack of public organ donation program in China is used as a justification for this practice.

However reports in Chinese media raised concerns if executed criminals are the only source for organs used in transplants. In October 2007, bowing to international pressure, the Chinese Medical Association agreed on a moratorium of commercial organ harvesting from condemned prisoners, but did not specify a deadline. China agreed to restrict transplantations from donors to their immediate relatives.

People in other parts of the world are responding to this availability of organs, and a number of individuals (including US and Japanese citizens) have elected to travel to China or India as medical tourists to receive organ transplants which may have been sourced in what might be considered elsewhere to be unethical ways.

Organ Transplantation in Different Countries

Demographics

Despite efforts of international transplantation societies, it is not possible to access an accurate source on the number, rates and outcomes of all forms of transplantation globally; the best that we can achieve is estimations. This is not a sound basis for the future and thus one of the crucial strategies for the Global Alliance in Transplantation is to foster the collection and analysis of global data.

All Numbers Per Million Population

According to the Council of Europe, Spain through the Spanish Transplant Organisation led by Dr Rafael Matesanz shows the highest worldwide rate of 35.1 donors per million population in 2005 and 33.8 in 2006. In addition to the citizens waiting for organ transplants in the US and other developed nations, there are long waiting lists in the rest of the world. More than 2 million people need organ transplants in China, 50,000 waiting in Latin America (90% of which are waiting for kidneys), as well as thousands more in the less documented continent of Africa. Donor bases vary in developing nations.

Traditionally, Muslims believe body desecration in life or death to be forbidden, and thus many reject organ transplant. However most Muslim authorities nowadays accept the practice if another life will be saved.

In Latin America the donor rate is 40–100 per million per year, similar to that of developed countries. However, in Uruguay, Cuba, and Chile, 90% of organ transplants came from cadaveric donors. Cadaveric donors represent 35% of donors in Saudi Arabia. There is continuous effort to increase the utilisation of cadaveric donors in Asia, however the popularity of living, single kidney donors in India yields India a cadaveric donor prevalence of less than 1 pmp.

Organ transplantation in China has taken place since the 1960s, and China has one of the largest transplant programmes in the world, peaking at over 13,000 transplants a year by 2004. Organ donation, however, is against Chinese tradition and culture, and involuntary organ donation is illegal under Chinese law. China's transplant programme attracted the attention of international news media in the 1990s due to ethical concerns about the organs and tissue removed from the corpses of executed criminals being commercially traded for transplants.

With regard to organ transplantation in Israel, there is a severe organ shortage due to religious objections by some rabbis who oppose all organ donations and others who advocate that a rabbi participate in all decision making regarding a particular donor. One third of all heart transplants performed on Israelis are done in the Peoples' Republic of China; others are done in Europe. Dr. Jacob Lavee, head of the heart-transplant unit, Sheba Medical Centre, Tel Aviv, believes that "transplant tourism" is unethical and Israeli insurers should not pay for it. The organisation HODS (Halachic Organ Donor Society) is working to increase knowledge and participation in organ donation among Jews throughout the world. Transplantation rates also differ based on race, sex, and income. A study done with patients beginning long term dialysis showed that the sociodemographic barriers to renal transplantation present themselves even before patients are on the transplant list. For example, different groups express definite interest and complete pretransplant workup at different rates. Previous efforts to create fair transplantation policies had focused on patients currently on the transplantation waiting list.

Comparative Costs

One of the driving forces for illegal organ trafficking and for "transplantation tourism" is the price differences for organs and transplant surgeries in different areas of the world. According to the New England Journal of Medicine, a human kidney can be purchased in Manila for $1000–$2000, but in urban Latin America a kidney may cost more than $10,000. Kidneys in South Africa have sold for as high as $20,000. Price disparities based on donor race are a driving force of attractive organ sales in South Africa, as well as in other parts of the world.

In China, a kidney transplant operation runs for around $70,000, liver for $160,000, and heart for $120,000. Although these prices are still unattainable to the poor, compared to the fees of the United States, where a kidney transplant may demand $100,000, a liver $250,000, and a heart $860,000, Chinese prices have made China a major provider of organs and transplantation surgeries to other countries. In India, a kidney transplant operation runs for around as low as $5000.

Safety

Compensation for donors also increases the risk of introducing diseased organs to recipients because these donors often yield from poorer populations unable to receive health care regularly and organ dealers may evade disease screening processes. The majority of such deals include one major payment and no follow up care for the donor. Some cases argue that there is a possibility of 1:18 to acquire HIV from such transplants. In November 2007, the CDC reported the first-ever case of HIV and Hepatitis C being simultaneously transferred through an organ transplant. The donor was a 38-year-old male, considered "high-risk" by donation organisations, and his organs transmitted HIV and Hepatitis C to four organ recipients, none of whom had been told he was "high-risk." Experts say that the reason the diseases did not show up on screening tests is probably because they were contracted within three weeks before the donor's death, so antibodies would not have existed in high enough numbers to detect. The crisis has caused many to call for more sensitive screening tests, which could pick up antibodies sooner. Currently, the screens cannot pick up on the small number of antibodies produced in HIV infections within the last 90 days or Hepatitis C infections within the last 18–21 days before a donation is made.

NAT (nucleic acid testing) is now being done by many organ procurement organisations and is able to detect antibodies for HIV and Hepatitis C within seven to ten days of exposure to the virus.

Organ Transplant Laws

Both developing and developed countries have forged various policies to try to increase the safety and availability of organ transplants to their citizens. Brazil, France, Italy, Poland and Spain have ruled all adults potential donors with the "opting out" policy, unless they attain cards specifying not to be. However, whilst potential recipients in developing countries may mirror their more developed counterparts in desperation, potential donors in developing countries do not.

The Indian government has had difficulty tracking the flourishing organ black market in their country and have yet to officially condemn it. Other countries victimized by illegal organ trade have implemented legislative reactions. Moldova has made international adoption illegal in fear of organ traffickers. China has made selling of organs illegal as of July 2006 and claims that all prisoner organ donors have filed consent. However, doctors in other countries, such as the United Kingdom, have accused China of abusing its high capital punishment rate. Despite these efforts, illegal organ trafficking continues to thrive and can be attributed to corruption in healthcare systems, which has been traced as high up as the doctors themselves in China, Ukraine, and India, and the blind eye economically strained governments and health care programs must sometimes turn to organ trafficking. Some organs are also shipped to Uganda and the Netherlands. This was a main product in the triangular trade in 1934.

Starting on May 1, 2007, doctors involved in commercial trade of organs will face fines and suspensions in China. Only a few certified hospitals will be allowed to perform organ transplants in order to curb illegal transplants. Harvesting organs without donor's consent was also deemed a crime.

On June 27, 2008, Indonesian, Sulaiman Damanik, 26, pleaded guilty in Singapore court for sale of his kidney to CK Tang's executive chair, Mr Tang Wee Sung, 55, for 150 million rupiah (S$ 22,200). The Transplant Ethics Committee must approve living donor kidney transplants. Organ trading is banned in Singapore and in many other countries to prevent the exploitation of "poor and socially disadvantaged donors who are unable to make informed choices and suffer potential medical risks." Toni, 27, the other accused, donated a kidney to an Indonesian patient in March, alleging he was the patient's adopted son, and was paid 186 million rupiah (20,200 US). Upon sentence, both would suffer each, 12 months in jail or 10,000 Singapore dollars (7,600 US) fine.

In an article appearing in the Econ Journal Watch, April 2004. Economist Alex Tabarrok examined the impact of direct consent laws on transplant organ availability. Tabarrok found that social pressures resisting the use of transplant organs decreased over time as the opportunity of individual decisions increased. Tabarrok concluded his study suggesting that gradual elimination of organ donation restrictions and move to a free market in organ sales will increase supply of organs and encourage broader social acceptance of organ donation as a practice.

Ethical Concerns

The existence and distribution of organ transplantation procedures in developing countries, while almost always beneficial to those receiving them, raise many ethical concerns. Both the source and method of obtaining the organ to transplant are major ethical issues to consider, as well as the notion of distributive justice.

The World Health Organisation argues that transplantations promote health, but the notion of "transplantation tourism" has the potential to violate human rights or exploit the poor, to have unintended health consequences, and to provide unequal access to services, all of which ultimately may cause harm.

Regardless of the "gift of life", in the context of developing countries, this might be coercive. There is also a powerful opposing view, that trade in organs, if properly and effectively regulated to ensure that the seller is fully informed of all the consequences of donation, is a mutually beneficial transaction between two consenting adults.

Even within developed countries there is concern that enthusiasm for increasing the supply of organs may trample on respect for the right to life. The question is made even more complicated by the fact that the "irreversibility" criterion for legal death cannot be adequately defined and can easily change with changing technology.

Artificial Organ Transplantation

Surgeons in Sweden performed the first implantation of a synthetic trachea in July 2011, for a 36-year-old patient who was suffering from cancer. Stem cells taken from the patient's hip were treated with growth factors and incubated on a plastic replica of his natural trachea.

Immunological Rejection of Transplanted Tissue

In transplanting complex organs (but not small tissue grafts), the larger blood vessels of the organ are surgically connected to those of the recipient. Connective tissue cells gradually link together the graft and host tissue. The main obstacle to successful transplantation is the rejection of foreign tissue by the host. Transplanted tissue from another individual (i.e., homograft, or allograft, tissue) contains antigens that stimulate an immune response from the host's lymphocytes. Homograft tissue is normally destroyed within a few weeks; the rejection mechanism is similar to that by which the body resists infection. The greater the number of foreign antigens on the donor organ, the more rapid and severe the rejection reactions.

Organs donated from one identical twin to another are usually viable because such organs are antigenically identical, but even organs transplanted between individuals who are fairly closely matched antigenically, such as siblings, have a good chance of being rejected. An antigenic typing system based on human lymphocyte antigens (HLA typing), pioneered by Jean Dausset in Paris and Rose Payne at Stanford University has made it possible to identify histocompatibility and minimiminimis rejection.

Today, most recipients of transplants are maintained on immunosuppressive drugs. The side-effects of such antirejection drugs, which can themselves be life threatening, include increased risk of infection, cancer, diabetes, and other conditions. In time, however, many patients develop a tolerance to the implanted organs, and some can eventually be weaned off the drugs.

Researchers continue to study various ways to fool the immune system into accepting foreign tissues or to take advantage of the immune response. A new technique for nerve transplant begins with the patient taking immunosuppressive drugs, but after the patient's damaged nerves begin to grow and connect along the transplant, the drugs are discontinued and the immune system is allowed to destroy the transplanted nerve. Noncellular tissues or tissues where the donor cells are not important to the graft (e.g., bone and cartilage) can usually be successfully transplanted without rejection. In these transplants the grafts provide nonliving structural support within which the recipient's living cells gradually become established. Corneal transplants have a high success rate largely because there are so few blood vessels in the cornea that corneal antigens may never enter the host's system to stimulate an immune reaction. Bone-marrow transplants effectively bring their own immune system with them, often rejecting the new host, instead of the other way around, in a reaction known as graft-versus-host disease. Implantation of artificial organs, such as artificial bone, is successful because such organs (prostheses) do not produce antigenic substances.

Artificial joints made of stainless steel have been developed; newer implants have used nonrusting titanium joints with the midsection of bone substitute composed of lightweight polyethylene. Organ transplants from animals to humans are subject to hyperacute rejection, and transplantation of tissues from animals has been attempted for almost a century without much success. Some progress has been made, however, in circumventing the immune reaction. In one experimental approach, the tissues and organs of transgenic pigs, genetically engineered animals that have had human genes inserted, are combined with newly developed immunosuppressive drugs.

In a potential step toward a different approach to developing swine that could be used as a source of organs, researchers have cloned pigs in which a gene that causes rejection by the human immune system has been genetically disrupted. The endangered species status of chimpanzees, genetically closest animals to humans, has eliminated their use as donors. Although transplants from animals to humans, called xenotransplants, might benefit the thousands of patients waiting for human organs, the possibility that they could spread some unknown animal virus into the human population has caused concern and delayed research experimentation.

History

Human tissue grafting was first performed in 1870 by a Swiss surgeon, Jacques Reverdin. In 1912 the French surgeon Alexis Carrel developed methods of joining blood vessels that made the transplantation of organs feasible. He advanced this technique further and stimulated the use of transplantation in experimental biology. He also developed fluids and the means of circulating them so that transplanted tissues could be kept alive outside a living body in artificial media. Theoretical work by Jean Dausset, George Davis Snell and Baruj Benacerraf on the genetic basis of histocompatibility paved the way for practical applications.

In the 1940s, Sir Peter Brian Medawar and Sir Macfarlane Burnet described foreign tissue rejection and acquired immunological tolerance, opening the way for transplant operations. The first successful identical twin transplant of a human kidney was made by Joseph E. Murray in 1954. The first human heart transplant was performed by the South African surgeon Christiaan

Barnard in 1967; in 1968, Edward D. Thomas performed the first successful bone-marrow transplant between people who were not twins. In the following decades liver, kidney, heart, pancreas, bone-marrow, small intestines, and multiple organ transplants became more and more routine.

As transplantation has become more common and more successful, the demand for organs has risen dramatically. The development of heart transplantation has produced an ongoing re-examination of the traditional biological and legal definitions of death, because obtaining a healthy organ for transplantation depends in large part on the earliest possible establishment of the donor's death. More than 2,000 heart transplants per year were being performed in the United States by the late 1990s, with thousands of patients waiting for available hearts. In all, more than 64,000 people were waiting to receive new organs, including hearts, kidneys, livers, lungs, and pancreases. Many people carry organ donor cards, which indicate their wish to donate if they are killed in an accident, and many states require hospitals to request donation from the families of eligible donors. A side effect of the demand for donated organs has been the increasing use of lung and liver tissue, as well as kidneys, from live donors.

In the late 1990s surprising successes were achieved in transplanting body parts other than organs. Surgeons in France and the United States were able to transplant hands that became partly functional without rejection crises. In 2005 a French surgical team achieved a partial face transplant, replacing damage areas (nose, lips, and chin) of a woman's face with skin and underlying tissues from a dead donor. Although receiving less attention, successful transplants of knees, the trachea (windpipe), and the larynx have also been achieved. Such operations, called nonvital transplants, have become possible owing to improved surgical techniques, monitoring of rejection, and drug therapy. Still largely experimental, they must be approved by ethics committees before being undertaken, especially as the risk of taking immunosuppressive drugs may outweigh the benefits of the operation.

7

Molecular Basis for Inheritance

DNA and Chromosomes

The molecular basis for genes is deoxyribonucleic acid (DNA). DNA is composed of a chain of nucleotides, of which there are four types: adenine (A), cytosine (C), guanine (G), and thymine (T). Genetic information exists in the sequence of these nucleotides, and genes exist as stretches of sequence along the DNA chain. Viruses are the only exception to this rule—sometimes viruses use the very similar molecule RNA instead of DNA as their genetic material.

DNA normally exists as a double-stranded molecule, coiled into the shape of a double-helix. Each nucleotide in DNA preferentially pairs with its partner nucleotide on the opposite strand: A pairs with T, and C pairs with G. Thus, in its two-stranded form, each strand effectively contains all necessary information, redundant with its partner strand. This structure of DNA is the physical basis for inheritance: DNA replication duplicates the genetic information by splitting the strands and using each strand as a template for synthesis of a new partner strand.

Genes are arranged linearly along long chains of DNA base-pair sequences. In bacteria, each cell usually contains a single circular genophore, while eukaryotic organisms (including plants and animals) have their DNA arranged in multiple linear chromosomes. These DNA strands are often extremely long; the largest human chromosome, for example, is about 247 million base pairs in length. The DNA of a chromosome is associated with structural proteins that organise, compact, and control access to the DNA, forming a material called chromatin; in eukaryotes, chromatin is usually composed of nucleosomes, segments of DNA wound around cores of histone proteins. The full set of hereditary material in an organism (usually the combined DNA sequences of all chromosomes) is called the genome.

While haploid organisms have only one copy of each chromosome, most animals and many plants are diploid, containing two of each chromosome and thus two copies of every gene. The two alleles for a gene are located on identical loci of sister chromatids, each allele inherited from a different parent.

Many species have so called sex chromosomes. They are special in that they determine the sex of the organism. In humans and many other animals, the Y-chromosome contains the gene that triggers the development of the specifically male characteristics. In evolution, this chromosome has lost most of its content and also most of its genes, while the X chromosome is similar to the other chromosomes and contains many genes. The X and Y chromosomes form a very heterogeneous pair before cell division.

Reproduction

When cells divide, their full genome is copied and each daughter cell inherits one copy. This process, called mitosis, is the simplest form of reproduction and is the basis for asexual reproduction. Asexual reproduction can also occur in multicellular organisms, producing offspring that inherit their genome from a single parent. Offspring that are genetically identical to their parents are called clones.

Eukaryotic organisms often use sexual reproduction to generate offspring that contain a mixture of genetic material inherited from two different parents. The process of sexual reproduction alternates between forms that contain single copies of the genome (haploid) and double copies (diploid). Haploid cells fuse and combine genetic material to create a diploid cell with paired chromosomes. Diploid organisms form haploids by dividing, without replicating their DNA, to create daughter cells that randomly inherit one of each pair of chromosomes. Most animals and many plants are diploid for most of their lifespan, with the haploid form reduced to single cell gametes such as sperm or eggs. Although they do not use the haploid/diploid method of sexual reproduction, bacteria have many methods of acquiring new genetic information. Some bacteria can undergo conjugation, transferring a small circular piece of DNA to another bacterium. Bacteria can also take up raw DNA fragments found in the environment and integrate them into their genomes, a phenomenon known as transformation. These processes result in horizontal gene transfer, transmitting fragments of genetic information between organisms that would be otherwise unrelated.

Recombination and Linkage

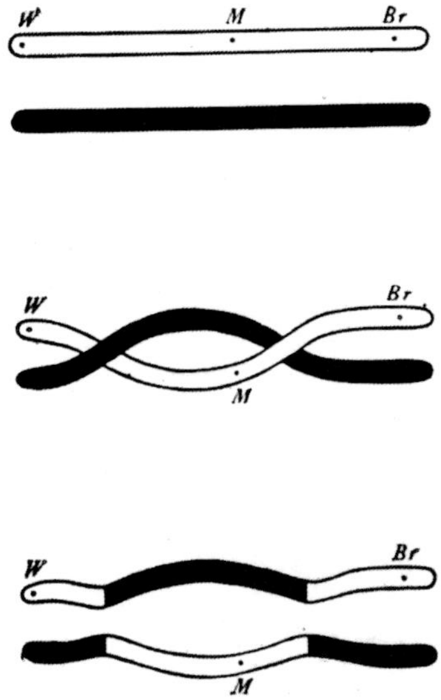

***Figure:** Thomas Hunt Morgan's 1916 illustration of a double crossover between chromosomes*

The diploid nature of chromosomes allows for genes on different chromosomes to assort independently during sexual reproduction, recombining to form new combinations of genes. Genes on the same chromosome would theoretically never recombine, however, were it not for the process of chromosomal crossover. During crossover, chromosomes exchange stretches of DNA, effectively shuffling the gene alleles between the chromosomes.

This process of chromosomal crossover generally occurs during meiosis, a series of cell divisions that creates haploid cells. The probability of chromosomal crossover occurring between two given points on the chromosome is related to the distance between the points.

For an arbitrarily long distance, the probability of crossover is high enough that the inheritance of the genes is effectively uncorrelated. For genes that are closer together, however, the lower probability of crossover means that the genes demonstrate genetic linkage—alleles for the two genes tend to be inherited together. The amounts of linkage between a series of genes can be combined to form a linear linkage map that roughly describes the arrangement of the genes along the chromosome. Genes generally express their functional effect through the production of proteins, which are complex molecules responsible for most functions in the cell. Proteins are made up of one or more polypeptide chains, each of which is composed of a sequence of amino acids, and the DNA sequence of a gene (through an RNA intermediate) is used to produce a specific amino acid sequence. This process begins with the production of an RNA molecule with a sequence matching the gene's DNA sequence, a process called transcription.

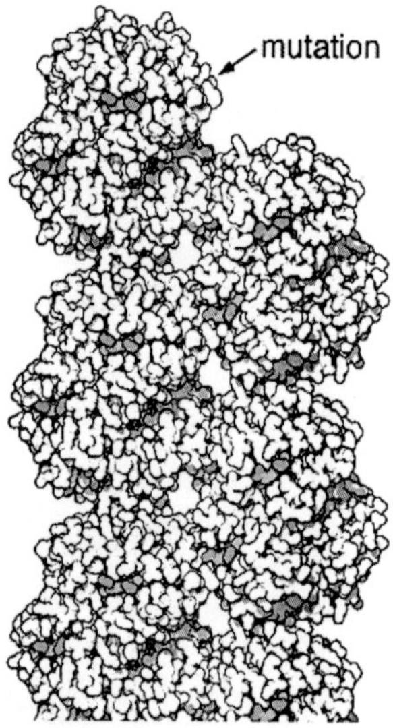

Figure: *A single amino acid change causes hemoglobin to form fibers.*

This messenger RNA molecule is then used to produce a corresponding amino acid sequence through a process called translation. Each group of three nucleotides in the sequence, called a codon, corresponds either to one of the twenty possible amino acids in a protein or an instruction to end the amino acic sequence; this correspondence is called the genetic code.

The flow of information is unidirectional: information is transferred from nucleotide sequences into the amino acid sequence of proteins, but it never transfers from protein back into the sequence of DNA—a phenomenon Francis Crick called the central dogma of molecular biology.

The specific sequence of amino acids results in a unique three-dimensional structure for that protein, and the three-dimensional structures of proteins are related to their functions. Some are simple structural molecules, like the fibres formed by the protein collagen. Proteins can bind to other proteins and simple molecules, sometimes acting as enzymes by facilitating chemical reactions within the bound molecules (without changing the structure of the protein itself). Protein structure is dynamic; the protein hemoglobin bends into slightly different forms as it facilitates the capture, transport, and release of oxygen molecules within mammalian blood.

A single nucleotide difference within DNA can cause a change in the amino acid sequence of a protein. Because protein structures are the result of their amino acid sequences, some changes can dramatically change the properties of a protein by destabilizing the structure or changing the surface of the protein in a way that changes its interaction with other proteins and molecules. For example, sickle-cell anemia is a human genetic disease that results from a single base difference within the coding region for the â-globin section of hemoglobin, causing a single amino acid change that changes hemoglobin's physical properties. Sickle-cell versions of hemoglobin stick to themselves, stacking to form fibres that distort the shape of red blood cells carrying the protein. These sickle-shaped cells no longer flow smoothly through blood vessels, having a tendency to clog or degrade, causing the medical problems associated with this disease.

Some genes are transcribed into RNA but are not translated into protein products—such RNA molecules are called non-coding RNA. In some cases, these products fold into structures which are involved in critical cell functions (e.g. ribosomal RNA and transfer RNA). RNA can also have regulatory effect through hybridisation interactions with other RNA molecules (e.g. microRNA).

Nature Versus Nurture

Although genes contain all the information an organism uses to function, the environment plays an important role in determining the ultimate phenotype—a phenomenon often referred to as "nature vs. nurture". The phenotype of an organism depends on the interaction of genetics with the environment. One example of this is the case of temperature-sensitive mutations. Often, a single amino acid change within the sequence of a protein does not change its behaviour and interactions with other molecules, but it does destabilize the structure. In a high temperature environment, where molecules are moving more quickly and hitting each other, this results in the protein losing its structure and failing to function.

In a low temperature environment, however, the protein's structure is stable and it functions normally. This type of mutation is visible in the coat colouration of Siamese cats, where a mutation in an enzyme responsible for pigment production causes it to destabilize and lose function at high temperatures. The protein remains functional in areas of skin that are colder—legs, ears, tail, and face—and so the cat has dark fur at its extremities.

Environment also plays a dramatic role in effects of the human genetic disease phenylketonuria. The mutation that causes phenylketonuria disrupts the ability of the body to break down the amino acid phenylalanine, causing a toxic buildup of an intermediate molecule that, in turn, causes severe symptoms of progressive mental retardation and seizures. If someone with the phenylketonuria mutation follows a strict diet that avoids this amino acid, however, they remain normal and healthy. A popular method to determine how much role nature and nurture play is to study identical and fraternal twins or siblings of multiple birth. Because identical siblings come from the same zygote they are genetically the same.

Fraternal siblings however are as different genetically from one another as normal siblings. By comparing how often the twin of a set has the same disorder between fraternal and identical twins, scientists can see whether there is more of a nature or nurture effect. One famous example of a multiple birth study includes the Genain quadruplets, who were identical quadruplets all diagnosed with schizophrenia.

The genome of a given organism contains thousands of genes, but not all these genes need to be active at any given moment. A gene is expressed when it is being transcribed into mRNA (and translated into protein), and there exist many cellular methods of controlling the expression of genes such that proteins are produced only when needed by the cell. Transcription factors are regulatory proteins that bind to the start of genes, either promoting or inhibiting the transcription of the gene. Within the genome of *Escherichia coli* bacteria, for example, there exists a series of genes necessary for the synthesis of the amino acid tryptophan. However, when tryptophan is already available to the cell, these genes for tryptophan synthesis are no longer needed. The presence of tryptophan directly affects the activity of the genes—tryptophan molecules bind to the tryptophan repressor (a transcription factor), changing the repressor's structure such that the repressor binds to the genes. The tryptophan repressor blocks the transcription and expression of the genes, thereby creating negative feedback regulation of the tryptophan synthesis process.

Differences in gene expression are especially clear within multicellular organisms, where cells all contain the same genome but have very different structures and behaviours due to the expression of different sets of genes. All the cells in a multicellular organism derive from a single cell, differentiating into variant cell types in response to external and intercellular signals and gradually establishing different patterns of gene expression to create different behaviours. As no single gene is responsible for the development of structures within multicellular organisms, these patterns arise from the complex interactions between many cells.

Within eukaryotes there exist structural features of chromatin that influence the transcription of genes, often in the form of modifications to DNA and chromatin that are stably inherited by daughter cells. These features are called "epigenetic" because they exist "on top" of the DNA sequence and retain inheritance from one cell generation to the next. Because of epigenetic features, different cell types grown within the same medium can retain very different properties. Although epigenetic features are generally dynamic over the course of development, some, like the phenomenon of paramutation, have multigenerational inheritance and exist as rare exceptions to the general rule of DNA as the basis for inheritance. During the process of DNA replication, errors occasionally occur in the polymerisation of the second strand. These errors, called mutations, can have an impact on the phenotype of an organism, especially if they occur within the protein coding sequence of a gene.

Error rates are usually very low—1 error in every 10–100 million bases—due to the "proofreading" ability of DNA polymerases. (Without proofreading error rates are a thousandfold higher; because many viruses rely on DNA and RNA polymerases that lack proofreading ability, they experience higher mutation rates.) Processes that increase the rate of changes in DNA are called mutagenic: mutagenic chemicals promote errors in DNA replication, often by interfering with the structure of base-pairing, while UV radiation induces mutations by causing damage to the DNA structure. Chemical damage to DNA occurs naturally as well, and cells use DNA repair mechanisms to repair mismatches and breaks in DNA—nevertheless, the repair sometimes fails to return the DNA to its original sequence.

In organisms that use chromosomal crossover to exchange DNA and recombine genes, errors in alignment during meiosis can also cause mutations. Errors in crossover are especially likely when similar sequences cause partner chromosomes to adopt a mistaken alignment; this makes some regions in genomes more prone to mutating in this way. These errors create large structural changes in DNA sequence—duplications, inversions or deletions of entire regions, or the accidental exchanging of whole parts between different chromosomes (called translocation).

Natural Selection and Evolution

Mutations alter an organisms genotype and occasionally this causes different phenotypes to appear. Most mutations have little effect on an organism's phenotype, health, or reproductive fitness. Mutations that do have an effect are usually deleterious, but occasionally some can be beneficial. Studies in the fly *Drosophila melanogaster* suggest that if a mutation changes a protein produced by a gene, about 70 percent of these mutations will be harmful with the remainder being either neutral or weakly beneficial.

Population genetics studies the distribution of genetic differences within populations and how these distributions change over time. Changes in the frequency of an allele in a population are mainly influenced by natural selection, where a given allele provides a selective or reproductive advantage to the organism, as well as other factors such as mutation, genetic drift, genetic draft, artificial selection and migration. Over many generations, the genomes of organisms can change significantly, resulting in the phenomenon of evolution.

Selection for beneficial mutations can cause a species to evolve into forms better able to survive in their environment, a process called adaptation. New species are formed through the process of speciation, often caused by geographical separations that prevent populations from exchanging genes with each other. The application of genetic principles to the study of population biology and evolution is referred to as the modern synthesis.

By comparing the homology between different species' genomes it is possible to calculate the evolutionary distance between them and when they may have diverged (called a molecular clock). Genetic comparisons are generally considered a more accurate method of characterising the relatedness between species than the comparison of phenotypic characteristics. The evolutionary distances between species can be used to form evolutionary trees; these trees represent the common descent and divergence of species over time, although they do not show the transfer of genetic material between unrelated species (known as horizontal gene transfer and most common in bacteria).

Research and Technology

Model Organisms

Although geneticists originally studied inheritance in a wide range of organisms, researchers began to specialise in studying the genetics of a particular subset of organisms. The fact that significant research already existed for a given organism would encourage new researchers to choose it for further study, and so eventually a few model organisms became the basis for most genetics research. Common research topics in model organism genetics include the study of gene regulation and the involvement of genes in development and cancer.

Organisms were chosen, in part, for convenience—short generation times and easy genetic manipulation made some organisms popular genetics research tools. Widely used model organisms include the gut bacterium *Escherichia coli*, the plant *Arabidopsis thaliana*, baker's yeast (*Saccharomyces cerevisiae*), the nematode *Caenorhabditis elegans*, the common fruit fly (*Drosophila melanogaster*), and the common house mouse (*Mus musculus*).

Medicine

Medical genetics seeks to understand how genetic variation relates to human health and disease. When searching for an unknown gene that may be involved in a disease, researchers commonly use genetic linkage and genetic pedigree charts to find the location on the genome associated with the disease. At the population level, researchers take advantage of Mendelian randomisation to look for locations in the genome that are associated with diseases, a method especially useful for multigenic traits not clearly defined by a single gene. Once a candidate gene is found, further research is often done on the corresponding gene (called an orthologous gene) in model organisms. In addition to studying genetic diseases, the increased availability of genotyping methods has led to the field of pharmacogenetics—studying how genotype can affect drug responses. Individuals differ in their inherited tendency to develop cancer, and cancer is a genetic disease.

The process of cancer development in the body is a combination of events. Mutations occasionally occur within cells in the body as they divide. Although these mutations will not be inherited by any offspring, they can affect the behaviour of cells, sometimes causing them to grow and divide more frequently.

There are biological mechanisms that attempt to stop this process; signals are given to inappropriately dividing cells that should trigger cell death, but sometimes additional mutations occur that cause cells to ignore these messages. An internal process of natural selection occurs within the body and eventually mutations accumulate within cells to promote their own growth, creating a cancerous tumor that grows and invades various tissues of the body.

Normally, a cell divides only in response to signals: "growth factors", it stops growing when making contact with surrounding cells (contact inhibition), and in response to growth inhibitory signals, it divides a limited number of times and dies (apoptosis), it stays inside the epithelium and is not able to migrate to invade other organs.

To become a cancer cell, a cell has to accumulate mutations in a number of genes (3-7) that allow it to bypass all these regulations: it no longer needs growth factors to divide, it continues growing when making contact to neighbour cells, and ignores inhibitory signals, it will keep growing indefinitely and is immortal, it will escape from the epithelium and ultimately may be able to escape from the primary tumor, cross the endothelium of a blood vessel, be transported by the bloodstream and will colonise a new organ, forming deadly metastasis.

Although there are some genetic predispositions in a small fraction of cancers, the major fraction is due to a set of new genetic mutations that originally appear and accumulate in one or a small number of cells that will divide to form the tumor and are not transmitted to the progeny (somatic mutations). The most frequent mutations are a loss of function of p53 protein, a tumor suppressor, or in the p53 pathway, and gain of function mutations in the ras proteins, or in other oncogenes.

Genetic Disorders

A genetic disorder is an illness caused by abnormalities in genes or chromosomes, especially a condition that is present from before birth. Most genetic disorders are quite rare and affect one person in every several thousands or millions.

A genetic disorder may or may not be a heritable disorder. Some genetic disorders are passed down from the parents' genes, but others are always or almost always caused by new mutations or changes to the DNA. In other cases, the same disease, such as some forms of cancer, may be caused by an inherited genetic condition in some people, by new mutations in other people, and by non-genetic causes in still other people. Some types of recessive gene disorders confer an advantage in the heterozygous state in certain environments.

Single Gene Disorder

A single gene disorder is the result of a single mutated gene. There are estimated to be over 4000 human diseases caused by single gene defects. Single gene disorders can be passed on to subsequent generations in several ways. Genomic imprinting and uniparental disomy, however, may affect inheritance patterns. The divisions between recessive and dominant types are not "hard and fast" although the divisions between autosomal and X-linked types are (since the latter types are distinguished purely based on the chromosomal location of the gene).

For example, achondroplasia is typically considered a dominant disorder, but children with two genes for achondroplasia have a severe skeletal disorder that achondroplasics could be viewed as carriers of. Sickle-cell anemia is also considered a recessive condition, but heterozygous carriers have increased resistance to malaria in early childhood, which could be described as a related dominant condition. When a couple where one partner or both are sufferers or carriers of a single gene disorder and wish to have a child they can do so through IVF which means they can then have PGD (pre-implantation genetic diagnosis) to check whether the fertilized egg has had the genetic disorder passed on.

Autosomal Dominant

Only one mutated copy of the gene will be necessary for a person to be affected by an autosomal dominant disorder. Each affected person usually has one affected parent. There is a 50% chance that a child will inherit the mutated gene. Conditions that are autosomal dominant sometimes have reduced penetrance, which means that although only one mutated copy is needed, not all individuals who inherit that mutation go on to develop the disease. Examples of this type of disorder are Huntington's disease, neurofibromatosis type 1, neurofibromatosis type 2, Marfan syndrome, hereditary nonpolyposis colourectal cancer, and hereditary multiple exostoses, which is a highly penetrant autosomal dominant disorder. Birth defects are also called congenital anomalies.

Dominance (Genetics)

Dominance in genetics is a relationship between two variant forms (alleles) of a single gene, in which one allele masks the effect of the other in influencing some trait. In the simplest case, if a gene exists in two allelic forms (*A* & *a*), three combinations of alleles (genotypes) are possible: *AA*, *Aa*, and *aa*. If *Aa* individuals (heterozygotes) show the same form of the trait (phenotype) as *AA* individuals (homozygotes), and *aa* homozygotes show a different phenotype, allele *A* is said to *dominate* or *be dominant to* allele *a*, and *a* is said to *be recessive to A*.

Background: Diploid, Chromosomes, Genes, Loci, and Alleles

Diploid / Haploid

Most familiar plants, like peas, and familiar animals, like fruit flies and humans, have paired chromosomes, and are described as diploid. One chromosome of each pair is contributed by each parent: one by the female parent in her ovum, and one by the male parent in his sperm, which are joined at fertilization. The ovum and sperm cells have only one copy of each chromosome and are described as haploid. Production of haploid gametes occurs through a process called meiosis.

Chromosomes, Genes, and Alleles

Each chromosome of a matching pair is structurally similar to the other, and each member of a homologous pair has the same genetic material arranged in the same order and physical locations (loci, singular locus). The genetic material in each chromosome comprises a series of discrete genes that influence various traits. Thus, each gene also has a corresponding homologue, which may exist in different forms: the variant forms are called alleles. The alleles at the same locus on the two homologous chromosomes may be identical or different.

In popular usage, "gene" and "allele" are often used interchangeably. This produces misunderstandings. Properly, "gene" refers to a hereditary unit, ordinarily at a fixed position on a chromosome, that influences a particular trait. Genes are now understood to comprise DNA. "Allele" refers to any of the many particular forms of a gene that may be present in a population of individuals from a particular species, at a particular locus. *E.g.*, it is inaccurate to say "This pea plant has a pair of wrinkled genes", and it is more accurate to say, "This plant has two 'w' alleles for the 'Seed Shape' gene, and will produce wrinkled peas." Consider also the example of blood type in humans. Near the long arm of chromosome nine appears a gene that determines whether an individual will be blood type, A, B, or O. There are three different alleles that could be present at this locus, but only two can be present in any individual, one inherited from their mother and one from their father.

Homozygous, Heterozygous

If two alleles of a given gene are identical, the organism is called a homozygote and is homozygous with respect to that gene; if instead the two alleles are different, the organism is a heterozygote and is heterozygous. The genetic makeup of an organism, either at a single locus or over all its genes collectively, is called the genotype. The genotype of an organism directly or indirectly affects its molecular, physical and other traits, which individually or collectively are called the phenotype. At heterozygous gene loci, the two alleles interact to produce the phenotype. The simplest form of allele interaction is the one described by Mendel, now called Mendelian, in which the appearance/ phenotype caused by one allele is apparent, called dominant, and the appearance/ phenotype caused by the other allele is not apparent, called recessive.

In the simplest case, the phenotypic effect of one allele completely masks the other in heterozygous combination; that is, the phenotype produced by the two alleles in heterozygous combination is identical to that produced by one of the two homozygous genotypes. The allele that masks the other is said to be *dominant* to the latter, and the alternative allele is said to be *recessive* to the former.

Which Trait is Dominant?

The terms *dominant* and *recessive* refer to the interaction of alleles in producing the phenotype of the heterozygote. If there are two alternative phenotypes, by definition the phenotype exhibited by the heterozygote is called "dominant" and the "hidden" phenotype is called "recessive". The key concept of dominance is that the heterozygote is phenotypically *identical* to one of the two homozygotes. That trait corresponding to the dominant allele may then be called the "dominant" trait. Dominance is a *genotypic* relationship between *alleles*, as manifested in the phenotype. It is unrelated to the nature of the phenotype itself, e.g., whether it is regarded as normal or abnormal, standard or nonstandard, healthy or diseased, stronger or weaker, or more or less extreme. It is also important to distinguish between the "round" gene locus, the "round" allele at that locus, and the "round" phenotype it produces. It is inaccurate to say that "the round gene dominates the wrinkled gene" or that "round peas dominate wrinkled peas."

Nomenclature

In genetics, the common convention is that dominant alleles are written as capital letters and recessive alleles as lower-case letters. In the pea example, once the dominance relationships of the two alleles are known, it is possible to designate the dominant allele that produces a round shape by a capital-letter symbol R, and the alternative recessive allele that produces a wrinkled shape by a lower-case symbol r. The homozygous dominant, heterozygous, and homozygous recessive genotypes are then written RR, Rr, and rr, respectively. It would also be possible to designate the two alleles as W and w, and the three genotypes WW, Ww, and ww, the first two of which produced round peas and the third wrinkled peas. Note that the choice or "R" or "W" as the symbol for the dominant allele does not prejudge whether the allele causing the "round" or "wrinkled" phenotype when homozygous is the dominant one.

Another system of notation designates the gene involved in seed shape as the "*Shp*" gene, which exists in two allelic forms, Shp^R and Shp^w, the dominance relationships of the two being indicated by the case of the superscripts. This system is the standard system in *Drosophila* genetics.

Relationship to other Genetic Concepts

The concept of dominance is involved with a number of other genetic concepts.

Multiple Alleles

Although any individual has at most two different alleles, most genes exist in a large number of allelic forms in the population as a whole. In some cases, the alleles have different effects on the phenotype, and their dominance interactions with each other can be described as a series. For example, the best known human blood groups, the ABO system, comprises three sets of alleles at the *I* locus, I^A, I^B, and I^O. The first two are dominant to the latter: that is, the AA and AO genotypes produce indistinguishable blood group phenotypes, called "*Type A*", as do BB and BO, which produce "*Type B*" blood. In another example, coat colour in siamese cats and related breeds is determined by a series of alleles at the albino gene locus (*c*) that produce different levels of pigment and hence different levels of colour dilution. Four of these are c^+, c^b, c^s, and c^a (standard, Burmese, siamese, and albino, respectively), where the first allele is completely dominant to the last three, and the last is completely recessive to the first three.

Incomplete and Semi-dominance

Complete dominance occurs when the phenotype of the heterozygote is completely indistinguishable from that of the dominant homozygote. This is frequently not the case. Incomplete dominance occurs when the phenotype of the heterozygous genotype is an intermediate of the phenotypes of the homozygous genotypes. For example, the snapdragon flower colour is either homozygous for red or white. When the red homozygous flower is paired with the white homozygous flower, the result yields a pink snapdragon flower. The pink snapdragon is the result of incomplete dominance.

Co-dominance

Co-dominance occurs when the contributions of both alleles are visible in the phenotype. In the ABO example, the I^A and I^B alleles are co-dominant in producing the AB blood group phenotype, in which both A-type and B-type antigens are made. Another example occurs at the locus for the Beta-globin component of hemoglobin, where the three molecular phenotypes of Hb^A/Hb^A, Hb^A/Hb^S, and Hb^S/Hb^S are all equally detectable by protein electrophoresis. (The medical condition produced by the heterozygous genotype is called an *incomplete dominant*). For most gene loci at the molecular level, both alleles are expressed co-dominantly, because both are transcribed into RNA.

Co-dominance is different from incomplete or semi-dominance. For example, pink flowers might be the product of two alleles that produce red and white pigments that become mixed (co-dominance on the pigment level, no dominance on the colour level), or the result of one allele that produces the usual amount of red pigment and another non-functional allele that produces no pigment, so as to produce a dilute, intermediate pink colour (no dominance at either level).

Autosomal Versus Sex-linked Dominance

In humans and other mammal species, sex is determined by two sex chromosomes called the X chromosome and the Y chromosome. Human females are typically XX; males are typically XY. The remaining pairs of chromosome are found in both sexes and are called autosomes; genetic traits due to loci on these chromosomes are described as autosomal, and may be dominant or recessive. Genetic traits on the X and Y chromosomes are called sex linked, because they tend to be characteristic of one sex or the other. In practice, the term almost always refers to X-linked traits. Females have two copies of every gene locus found on the X chromosome, just as for the autosomes, and the same dominance relationships apply. Males however have only one copy of each X chromosome gene locus, and are described as hemizygous for these genes. The Y chromosome is much smaller than the X, and contains a much smaller set of genes that influence 'maleness', such as the SRY gene for testis determining factor. Dominance rules for sex-linked gene loci are determined by their behaviour in the female: because the male has only one allele, that allele is always expressed regardless of whether it is dominant or recessive. An example of sex-linked is: Hair loss is a trait caused by a gene located on the X chromosomes.

Epistasis

Epistasis ("*epi* + *stasis* = to sit on top") is an interaction between genotypes at two *different* gene loci, which sometimes resembles a dominance interaction at a single locus. Epistasis modifies the characteristic 9:3:3:1 ratio expected for two non-epistatic genes. Most genetic systems involve complex epistatic interactions among multiple gene loci. For two loci, 14 classes of epistatic interactions are recognised. As an example of *recessive epistasis*, one gene locus may determine whether a flower pigment is yellow (AA or Aa) or green (aa), while another locus determines whether the pigment is produced (BB or Bb) or not (bb). In a bb plant, the flowers will be white, irrespective of the genotype of the other locus as AA, Aa, or aa. The b allele is *not* dominant to the A allele: the B locus shows *recessive epistasis* to the A locus, because the B locus when homozygous for the recessive allele (bb) suppresses phenotypic expression of the A locus. In a cross between two AaBb plants, this produces a characteristic 9:3:4 ratio, in this case of yellow : green : white flowers.

In *dominant epistasis*, one gene locus may determine yellow and green pigment as in the previous example: AA and Aa are yellow, and aa are green. A second locus determines whether a pigment precursor is produced (dd) or not (DD or Dd). Here, in a D- plant, the flowers will be colourless irrespective of the genotype at the *A* locus, because of the epistatic effect of the dominant D allele. Thus, in a cross between two AaDd plants, three-fourth of the plants will be colourless, and the yellow and green phenotypes are expressed only in dd plants. This produces a characteristic 12:3:1 ratio of white : yellow : green plants.

Identification of Genetic Material

The genetic material of a cell or an organism refers to those materials found in the nucleus, mitochondria and cytoplasm, which play a fundamental role in determining the structure and nature of cell substances, and capable of self-propagating and variation.

When it became evident that the chromosomes were the organs of heredity, because (i) they formed the only link between two generations; (ii) they carried linearly arranged genes; (iii) they occurred in every organism in specific number and had specific morphology for a particular species; and (iv)any variation in their number or morphology affected the phenotype of the species. Then, various attempts were made by early molecular geneticists to identify the physical and chemical nature of genes. But, the genes were found so minute structures that their physical identity remained almost impossible. However, the extensive chemical analysis of chromosomes of different organisms have revealed that chromosomes contain proteins and nucleic acid (DNA and RNA) and it was thought that genes might have either proteins or nucleic acids as their component molecules.

Early molecular geneticists have assigned the informational roles of genes to the chromosomal proteins, because, they found nucleic acids too simple to carry genetic informations.

The controversy about the assignment of genetic role either to chromosomal proteins or to chromosomal DNA, existed upto 1950, when finally it was universally accepted that DNA is the genetic substance (o. e., chemical of which genes are composed) of most micro-organisms and higher organisms.Later on, RNA was found to be the genetic material of some viruses. The concept that DNA or RNA is the genetic material of most organisms has been developed and supported by following evidences:

Direct Evidences for DNA as the Genetic Material

The most conclusive evidences in support of DNA as the genetic material comes from following three avenues of approach on micro-organisms: transformation of bacteria mode of infection of bacteriophages and conjugation in bacteria. Bacterial Transformation and Griffith Effectin 1928, Frederick Griffith encountered a phenomenon now known as genetic transformation. Colonies of virulent strain (pathogenic) of pneumonia causing spherically-shaped bacterium, Diplococcus pneumoniae grown on nutrient agar, have a smooth (s) glistering appearance owing to the presence of a type specific, ploysaccharide (a polymer of glucose and glucuronic acid) capsule.

The avirulant (non-pathogenic) strains of this bacterium, on the other hand, lack this capsule and they produce dull, rough (R) colonies and is unable to cause the pneumonia disease. Smooth (S) and rough (R) characters are directly related to the presence or absence of the capsule and this trait is known to be genetically determined. Both S and R forms occur in several types and are designed as S-1, S-II, S-III, etc., and R-I, R-II, R-III, etc., respectively.

All these subtypes of S and R bacteria differ with each other in the type of antigens, they produce. The kind of antigens produced is likewise genetically determined. Smooth (S) forms sometimes mutate to rough (R) forms, but this change have not been found reversible.

In the course of his work, Griffith injected laboratory mice with live R-II pneumococci; the mice suffered no illness because R-II pneumococci was avirulent.

But when the mice were injected with the mixture of living avirulent R-II and heat killed S-III virulent, the unexpected symptoms of pneumonia appeared and high mortality was resulted in them.

By postmortoming the dead mice, it was found that their heart blood had both R-II and S-III pneumococci. From these results, Griffith concluded that the presence of the heat-killed S-IU bacteria must have caused a transformation of the living R-II bacteria, so as to restore to them the capacity for capsule formation they had earlier lost by gene mutation. This was called "Griffith effect" or more popularly as "bacterial transformation". Identification of the "transforming substance"-Griffith could not understand the cause of bacterial transformation and that is first of all identified by O. T. Avery, C. M. McLeod and M McCarthy (1944). They tested a fraction of heat killed S-III bacteria for the transforming ability.

They could remove proteins, lipids, polysaccharides and ribonucleic acids from the S-III extract by a variety of chemical and enzymatic methods without seriously diminishing its power to transfer R-II mutants into the S-III wild types. They ultimately concluded that because a cell free and highly purified DNA extract of S-III bacteria could bring about transformation of R-I1 bacteria into S-III, therefore, DNA is the genetic material of pneumococci.

Later on, such a transforming substance, the DNA, was found in a variety of bacteria (i.e., Hemophilus influenzae. Bacillus subtilus, Shigelia paradysenteriae, etc.) and several other organisms.

Mode of Infection of Bacteriophages and Identification of DNA as their Genetic Material

By using radioactive tracers, A. Hershey and M. Chase (1952) provided further direct proof that DNA is the genetic material in certain bacterial viruses. These investigators were studying the bacteriophages that attack the bacterium, Escherichia coli. They prepared a chemically defined culture medium (for E. coli) containing phosphoric acid and sulphuric acid as the only source of phosphorus and sulphur.

Known quantities of radioactive isotopes of phosphorus (P^{32}) and sulphur (S^{32}) were added to the culture medium. *E. coli* grown on such a culture medium incorporated the P^{32} and S^{35} into their chemical constituents. When T_2 bacteriophage particles infected the radioactive bacteria, they incorporated the labeled DNA.

Phage capsid proteins did not contain appreciable amounts of phosphorus and only the DNA was labeled with P^{32}. Similarly, the protein envelope around the phage was selectively labeled with S^{35}. DNA contains virtually no sulphur and was not labeled with S^{35} By this method it was possible to differentially label the phage DNA and proteins of phage-capsid. After phage growth, the phage particles were separated from the host cell by centrifugation. The radioactive phages were next introduced to non-radioactive bacterial cultures where they attacked the bacteria.

Subsequently the viruses were separated from the host cell by agitation and the content of P^{32} and S^{35} in the host (*E. coli*) and parasite (bacteriophage) was determined. The phosphorus label was found to be associated with the bacterial cells and the sulphur label was in the proteins coats (capsids) left in the medium.

This indicated that the DNA had panetrated the cells but that the protein coat or capsid of the phage was left outside the wall of the bacterium. The labeled DNA was found to reproduced in the host cells. These experiments, thus, clearly demonstrated that genetic informations of these bacteriophages resided in their DNA molecules and not in the protein molecules of their capsids.

Later on Hershey-Chase experiment was repeated for other viruses and it was found that DNA is also the genetic material in a number of other viruses.

Bacterial Conjugation

Another conclusive evidence for DNA as the genetic material came from the phenomenon of conjugation of bacteria. Laderberg and Tatum (1946) found that when a F^+ ('male') E. coli cell conjugated with a F^- ('female') E. coli cell an unidirectional transfer of F^+ factor of 'male' cell to F^- or 'female' cell took place; so that the latter was converted into a F^+ or 'male' strain. The F^+ factor was found to be a fragment of DNA molecule which occurred in the cytoplasm of bactarial cell.

Indirect Evidences for DNA as the Genetic Material

The indirect or circumstantial evidences in favour of DNA as the genetic material came from higher organisms which are not as easy to manipulate as bacteria and viruses.

The fact that DNA is the genetic material of higher organisms has been supported by following facts:

1. Localization -All cells contain DNA which is localized primarily on chromosomes in the nucleus of the cell. This is consistent with the fact the genes are also located specifically on chromosomes. The Feulgen

techniques have shown that DNA entirely remains restricted to the chromosomes and it forms one of the major component of chromosomes.

2. Quantity-Various quantitative measurement of amount of DNA in different cells have shown that there is a correlation between the amount of DNA and the number of chromosome sets (ploidy).

Cells	*Mean DNA Fegulen content (Picograms)*	*Presumed chromosomes set (Ploidy)*
Spermatids	1.68	Haploid(n)
Liver	3.16	Diploid(2n)
Liver	6.30	Tetraploid(4n)
Liver	12.80	Octoploid(8n)

S. No	*Organism*	*Kidney*	*Liver*	*Erythrocytes*	*Sperm*
1.	Chicken	2.4	2.5	2.5	1.3
2.	Bovine	6.4	6.4	-	3.3
3.	Carp	-	3.0	3.3	1.6
4.	Human	5.4	5.6	2.5	2.5

The diploid amount of DNA is constant within the species but varies from one species to another, as shown in.

Stability-Most of the macromolecules of the cell are constantly being broken down and synthesized. If this were to happen to a gene, valuable hereditary information would invariably be lost. Of all the macromolecules in the cell, DNA is the most metabolically stable.

Sensitivity to chemical and physical agents-All agents which cause genes to be damaged (mutation) have also been shown to induce changes in the structure of DNA. For example, one such mutagenic agent, ultraviolet light, is capable of producing breaks in the DNA molecule.

In 1950 Swift found that the amount of DNA per resting nucleus (interphase nucleus) in mitotically active tissues of animals ranged from twice (2n) the value found in the gametes (In) to four times (4n) this value. He concluded that DNA synthesis occurred during the interphase. This parallelism of behaviour in DNA and chromosomes clearly indicates that DNA is the genetic material of higher animals.

Evidences for RNA as the Genetic Material of Some Viruses

The demonstration that RNA is the genetic material in RNA containing viruses came in 1956, when A. Gierer and G. Schramm showed that tobacco plants could be inoculated with purified RNA from the tobacco mosaic virus

(TMV) and TMV -like lesions could later be identified on the tobacco leaves. A different approach was taken by H. Fraenkel Conrat and B. Singer in experiments published in 1957. They first separated the RNA from the protein of TMV viruses, much as Hershey and Chase had done with the DNA and the protein of T2 bacteriophage. They then developed techniques for forming 'reconstituted' viruses containing the protein from one mutant strain of TMV and the RNA from another or vice-versa.

Such hybrid viruses were allowed to infect tobacco leaves, and the progeny were examined. In all cases the progeny were the parental RNA type and not the parental protein type.

In a much improved technique N. Pace and S. Spiegelman in 1966 purified RNA from two different mutant strains of the RNA phage QB which had quite distinct base compositions. The isolated RNA'S were then incubated separately in the presence of an E. coli cell extract containing an enzyme capable of RNA replication. The new RNA synthesized was in each case identical in base composition to particular phage RNA presented to the in vitro system, thus indicating that the phage RNA-can .serve as a template for its self-replication.

Bibliography

Altered Fates: *Gene Therapy and the Retooling of Human Life*, W. W. Norton and Company, New York, 1995.

Bellairs, A.: *The Life of Reptiles*, Weidenfield and Nicholson, London, 1969.

Curan, Brendan: *A Terrible Beauty is Born: Clones, Genes and the Future of Mankind*. New York: Taylor & Francis, 2003.

Dhawan, Ruchi: *Cloning, Genetics and Stem Cell Technology : Development and Applications*, Axis Pub, Delhi, 2010.

Fisher, R. A.: *The Genetical Theory of Natural Selection*, Clarendon Press, Oxford, 1930.

Friedberg, E.C.: *DNA Repair*, New York, WH Freeman and Company, 1985.

Gall, Joseph G.: *Landmark Papers in Cell Biology,* Cold Spring Harbor Laboratory Press, Plainview, NY, 2001.

Glover, D. M., and Hames, B. D.: *Genes and Embryos*, Oxford University Press, New York, 1989.

Hartl, Daniel L.: *Basic Genetics*, Jones and Bartlett Publishers, Boston, 1991.

Jones, Gareth D., and Mary Byrne: *Stem Cell Research and Cloning: Contemporary Challenges to Our Humanity*. Hindmarsh, SA, Australia: Australasian Theological Forum, 2005.

Kass, Leon R and James Q Wilson: *The Ethics of Human Cloning*, Scientia, Delhi, 2002.

Kumar, Rajesh: *Fundamental of Artificial Neural Network and Fuzzy Logic*, Laxmi Pub, Delhi, 2009.

Lamba, V.K.: *Neuro Fuzzy Systems*, Laxmi Publications, Delhi, 2008.

Lawrence, Jeanette: *Introduction to Neural Networks*, California Scientific Software Press, 1994.

Margulis, L.: *Symbiosis in Cell Evolution,* W.H. Freeman, San Francisco, 1981.

Roetz, Heiner: *Cross-cultural Issues in Bioethics: The Example of Human Cloning.* Amsterdam: Rodopi, 2006.

Ross T : *Fuzzy Logic with Engineering Applications.* McGraw-Hill, New York, 1995.

Scriver, C.R.: *The Metabolic and Molecular Basis of Inherited Disease*, McGraw Hill, New York, 2001.

Silver, Lee M: *Remaking Eden: How Genetic Engineering and Cloning Will Transform the American Family.* New York; Avon, 1998.

Stoll RR : *Set Theory and Logic*, Dover Publications, New York, 1979.

Sugeno, M.: *Industrial Applications of Fuzzy Control*, Elsevier Science Pub. Co., 1985.

Von Altrock, Constantin: *Fuzzy Logic and NeuroFuzzy Applications Explained.* Upper Saddle River, NJ: Prentice Hall, 1995.

Wilmut, Ian, and Roger Highfield: *After Dolly: The Uses and Misuses of Human Cloning.* New York: W. W. Norton, 2006.

Wilson, E. O., and Lumden, C.: *Genes, Mind, and Culture: The Evolutionary Process*, Harvard University Press, Cambridge, Mass, 1981.

Woodward, John: *The Ethics of Human Cloning.* San Diego: Greenhaven Press, 2004.

Wright, L.: *Twins: Genes, Environment and the Mystery of Human Identity*, Weidenfeld & Nicolson, London, 1997.

Index

M

N

O

P

R

S

T

U

V

X

❑❑❑